advancing learning, changing lives

Edexcel IGCSE
Human Biology

D1545287

Student Book

Phil Bradfield, Steve Potter

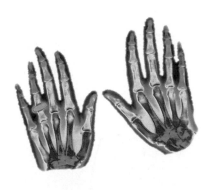

A PEARSON COMPANY

Published by Pearson Education Limited, a company incorporated in England and Wales, having its registered office at Edinburgh Gate, Harlow, Essex, CM20 2JE. Registered company number: 872828.

www.pearsonschoolsandfecolleges.co.uk

Edexcel is a registered trademark of Edexcel Limited

Text © Pearson Education Ltd 2010

First published 2010

15 14 13 12 11

10 9 8 7 6 5 4 3

ISBN 978 0 435044 13 8

Original design by Richard Ponsford and Creative Monkey

Original illustrations © Pearson Education Ltd 2010

Packaged and typeset by Naranco Design & Editorial / Kim Hubbeling

Edited by Anne Sweetmore

Proofread by Liz Jones

Illustrated by HL Studios, Bayard Christ and Cláudia Fonseca

Cover design by Creative Monkey

Cover photo © Jupiter Images / Brand X / Alamy

Index by Indexing Specialists (UK) Limited

Printed by Multivista Global Ltd

Acknowledgements

The author and publisher would like to thank the following individuals and organisations for permission to reproduce photographs:

(Key: b-bottom; c-centre; l-left; r-right; t-top)

Alamy Images: Chris Batson 198c, Deborah Davis 99, M. Scott Brauer 211, Young-Wolff Photography 155; Corbis: Lester V. Bergman 43, Bettmann 158, 193, Clouds Hill Imaging Ltd 150t, F.Carter Smith / Sygma 133, Image Source 84; DK Images: Steve Gorton 30; Victor Englebert: 29; FLPA Images of Nature: Konrad Wothe 217, Nigel Cattlin 198t;

Food Features: 204; Getty Images: 180b, Hulton Archive / Davies 195r, John Wang 183, John Kelly / Riser 100, Nancy R Cohen / Photodisc 71b, Stone / Terje Rakke 91; Holt Studios International Ltd: Nigel Cattlin 153t; http://www.un.org/Docs/sc/unsc_functions.html: Eskinder Debebe 113br; Natural History Museum Picture Library: 83; Panos Pictures: Liba Taylor 145t; Photolibrary.com: OSF 178, Howard Rice 164; Phototake, Inc: C. James Webb 195l, Dennis Kunkel Microscopy, Inc 2; Science Photo Library Ltd: 28, 139b, Michael Abbey 174r, AJ Photo / Hop Americain 115, Andy Crump, TDR, WHO 35b, Astrid & Hanns-Frieder Michler 92, 177, 198r, A. Barrington Brown 139t, Alex Bartel 65, Juergen Berger 176r, Biocosmos / Francis Leroy 122, Biology Media 63, 189, Biophoto Associates 36b, 37, Brad Nelson/Custom Medical Stock Photo 114t, BSIP 174cl, Conor Caffrey 27b, Gene Cox 96, Dept. of Clinical Cytogenetics, Addenbrookes Hospital 143l, 143r, Martin Dohrn 186b, Dr Gopal Murti 3t, 95, 150b, Dr Keith Wheeler 129, Dr Linda Stannard, UCT 180t, Dr M A Ansary 185, John Durham 145b, Eye of Science 21b, 47, 174cr, Steve Gschmeissner 12, GustoImages 113bl, Adam Hart-Davis 131b, 216t, J.C. REVY, ISM 3b, 36t, Jim Varney 131t, John Paul Kay, Peter Arnold INC 89, James King-Holmes 154, Kwangshin Kim 174l, Leonard Lessin 71t, 71c, Manfred Kage 188, Cordelia Molloy 93l, Moredun Animal Health Ltd 176l, NIBSC 181, OMIKRON 169b, Chris Priest 114b, Saturn Stills 93r, 192, Science Pictures Limited 153b, Science Pictures Ltd 151, St. Mary's Hospital Medical School 182, Zephyr 184; STILL Pictures The Whole Earth Photo Library: Ron Giling / Lineair 216b

All other images © Pearson Education 2010

Every effort has been made to contact copyright holders of material reproduced in this book. Any omissions will be rectified in subsequent printings if notice is given to the publishers.

Websites

The websites used in this book were correct and up to date at the time of publication. It is essential for tutors to preview each website before using it in class so as to ensure that the URL is still accurate, relevant and appropriate. We suggest that tutors bookmark useful websites and consider enabling students to access them through the school/college intranet.

Disclaimer

This material has been published on behalf of Edexcel and offers high-quality support for the delivery of Edexcel qualifications.

This does not mean that the material is essential to achieve any Edexcel qualification, nor does it mean that it is the only suitable material available to support any Edexcel qualification. Edexcel material will not be used verbatim in setting any Edexcel examination or assessment. Any resource lists produced by Edexcel shall include this and other appropriate resources.

Copies of official specifications for all Edexcel qualifications may be found on the Edexcel website: www.edexcel.com.

Contents

About this book

This book has several features to help you with IGCSE Human Biology.

Introduction
Each chapter has a short introduction to help you start thinking about the topic and let you know what is in the chapter.

End of Chapter Checklists
These lists summarise the material in the chapter. They could also help you to make revision notes because they form a list of things that you need to revise. (You need to check your specification to find out exactly what you need to know.)

Chapter 3: Food and Digestion

We need food for three main reasons:

- to supply us with a 'fuel' for energy
- to provide materials for growth and repair of tissues
- to help fight disease and keep our bodies healthy.

Food is essential for life. The nutrients obtained from it are used in many different ways by the body. This chapter looks at the different kinds of food, and how the food is broken down by the digestive system and absorbed into the blood, so that it can be carried to all the tissues of the body.

A balanced diet

The food that we eat is called our **diet**. No matter what you like to eat, if your body is to work properly and stay healthy, your diet must include five groups of food substances – **carbohydrates**, **lipids**, **proteins**, **minerals** and **vitamins** – as well as **water** and **fibre**. Food should provide you with all of these substances, but they must also be present in the *right* amounts. A diet that provides enough of these substances and in the correct proportions to keep you healthy is called a **balanced diet** (Figure 3.1). We will deal with each type of food in turn, to find out about its chemistry and the role that it plays in the body.

Figure 3.1 *A balanced diet contains all the types of food the body needs, in just the right amounts.*

Carbohydrates

Carbohydrates only make up about 5% of the mass of the human body, but they have a very important role. They are the body's main 'fuel' for supplying cells with energy. Cells release this energy by oxidising a sugar called **glucose**, in the process called cell respiration (see Chapter 1). Glucose and other sugars are one sort of carbohydrate.

Glucose is found naturally in many sweet-tasting foods, such as fruits and vegetables. Other foods contain different sugars, such as the fruit sugar called **fructose**, and the milk sugar, **lactose**. Ordinary table sugar, the sort some people put in their tea or coffee, is called **sucrose**. Sucrose is the main sugar that is

The chemical formula for glucose is $C_6H_{12}O_6$. Like all carbohydrates, glucose contains only the elements carbon, hydrogen and oxygen. The 'carbo' part of the name refers to carbon, and the 'hydrate' part refers to the fact that the hydrogen and oxygen atoms are in the ratio two to one, as in water (H_2O).

End of Chapter Checklist

You should now be able to:

- ✓ understand that division of a diploid cell by mitosis produces two cells that contain identical sets of chromosomes
- ✓ understand a simple outline description of the four stages of mitosis
- ✓ recall that the diploid number of chromosomes in humans is 46 and the haploid number is 23
- ✓ understand that division of a cell by meiosis produces four cells, each with half the number of chromosomes, and that this results in the formation of genetically different haploid gametes
- ✓ understand that mitosis occurs during growth, repair, asexual reproduction and cloning
- ✓ understand that variation within a species can be genetic, environmental or a combination of both

Questions

1 Cells can divide by mitosis or by meiosis.

  a) Give one similarity and two differences between the two processes.

  b) Do cancer cells divide by mitosis or meiosis? Explain your answer.

  c) Why is meiosis sometimes called reduction division?

2 Some cells divide by mitosis, others divide by meiosis. For each of the following examples, say whether mitosis or meiosis is involved. In each case, give a reason for your answers.

  a) Cells in the testes dividing to form sperm.

  b) Cells in the lining of the small intestine dividing to replace cells that have been lost.

  c) Cells in the bone marrow dividing to form red blood cells and white blood cells.

  d) A zygote dividing to form an embryo.

3 Variation in organisms can be caused by the environment as well as by the genes they inherit. For each of the following examples, state whether the variation described is likely to be genetic, environmental or both. In each case, give a reason for your answers.

  a) Humans have brown, blue or green eyes.

  b) Half the human population is male, half is female.

  c) Cuttings of hydrangea plants grown in soils with different pH values develop flowers with slightly different colours.

  d) People in some families are more at risk of heart disease than people in other families. However, not every member of the 'high risk' families has a heart attack and some members of the 'low risk' families do.

4 In an investigation into mitosis, the distance between a chromosome and the pole (end) of a cell was measured. The graph shows the result of the investigation.

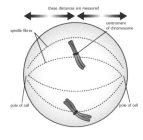

Margin Boxes
The boxes in the margin give you extra help or information. They might explain something in a little more detail or guide you to linked topics in other parts of the book.

Questions
There are short questions at the end of each chapter. These help you to test your understanding of the material from the chapter.

Chapter 1: Life Processes

The human body is composed of countless millions of units called **cells**. In an animal like a human there are many different types of cells, with different structures. They are specialised so that they can carry out particular functions in the body. Despite all the differences, there are basic features that are the same in most cells.

The cells of humans and other organisms share common features. In this chapter you will read about these features and look at some of the processes that keep cells alive.

Cell structure

For over 160 years scientists have known that animals and plants are made from cells. All cells contain some common parts, such as the nucleus, cytoplasm and cell membrane. Some cells have structures missing, for instance red blood cells lack a nucleus, which is unusual. The first chapter in a biology textbook usually shows diagrams of 'typical' plant and animal cells. In fact, there is really no such thing as a 'typical' cell. Humans, for example, are composed of hundreds of different kinds of cells from nerve cells to blood cells, skin cells to liver cells. What we really mean by a 'typical' cell is a general diagram that shows all the features that you might find in most cells, without them being too specialised. Figure 1.1 shows the features you would expect to see in many animal and plant cells. However, not all these are present in all cells – the parts of a plant which are not green do not have chloroplasts, for example.

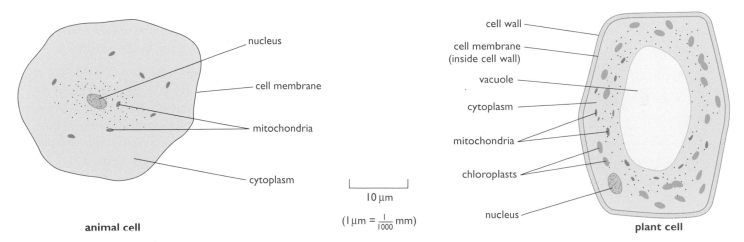

Figure 1.1 *The structure of a 'typical' animal and plant cell.*

The living material that makes up a cell is called **cytoplasm**. It has a texture rather like sloppy jelly, in other words somewhere between a solid and a liquid. Unlike a jelly, it is not made of one substance but is a complex material made of many different structures. You can't see many of these structures under an ordinary light microscope. An electron microscope has a much higher magnification, and can show the details of these structures, which are called **organelles** (Figure 1.2).

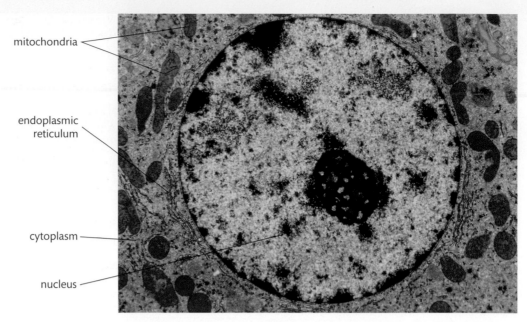

mitochondria

endoplasmic
reticulum

cytoplasm

nucleus

Figure 1.2 *The organelles in a cell can be seen using an electron microscope.*

The largest organelle in the cell is the **nucleus**. Nearly all cells have a nucleus, with a few exceptions, such as red blood cells. The nucleus controls the activities of the cell. It contains **chromosomes** (46 in human cells) which carry the genetic material, or **genes**. You will find out much more about genes and inheritance later in the book. Genes control the activities in the cell by determining which proteins the cell can make. One very important group of proteins found in cells is **enzymes** (see below). Enzymes control chemical reactions that go on in the cytoplasm.

All cells are surrounded by a **cell surface membrane** (often simply called the cell membrane). This is a thin layer like a 'skin' on the surface of the cell. It forms a boundary between the cytoplasm of the cell and the outside. However, it is not a complete barrier. Some chemicals can pass into the cell and others can pass out (the membrane is **permeable** to them). In fact, the cell membrane *controls* which substances pass in either direction. We say that it is **selectively** permeable.

There are other membranes inside a cell. Throughout the cytoplasm is a network of membranes called the **endoplasmic reticulum**. In places the endoplasmic reticulum is covered with minute granules called **ribosomes**. These are the organelles where proteins are assembled.

One organelle that is found in the cytoplasm of all living cells is the **mitochondrion** (plural **mitochondria**). There are many mitochondria in cells that need a lot of energy, such as muscle or nerve cells. This gives us a clue to the role of mitochondria. They carry out some of the reactions of **respiration** (see page 6) to release energy that the cell can use. In fact, most of the energy from respiration is released in the mitochondria.

All of the structures we have seen so far are found in both animal and plant cells. However, some structures are only ever found in plant cells. There are three in particular – the cell wall, a permanent vacuole and chloroplasts.

The **cell wall** is a layer of non-living material that is found outside the cell membrane of plant cells. It is made mainly of a carbohydrate called **cellulose**, although other chemicals may be added to the wall in some cells. Cellulose is a tough material that helps the cell keep its shape. This is why plant cells have a fairly fixed shape. Animal cells, which lack a cell wall, tend to be more variable in shape.

Plant cells often have a large central space surrounded by a membrane, called a **vacuole**. This vacuole is a permanent feature of the cell. It is filled with a watery liquid called **cell sap**, a store of dissolved sugars, mineral ions and other solutes. Animal cells can have small vacuoles, but they are only temporary structures.

Cells of the green parts of plants, especially the leaves, have another very important organelle, the **chloroplast**. Chloroplasts absorb light energy to make food in the process of photosynthesis. The chloroplasts are green because they contain a green pigment called **chlorophyll**.

Figure 1.3 shows some animal and plant cells seen through the light microscope.

Enzymes: controlling reactions in the cell

The chemical reactions that go on in a cell are controlled by a group of proteins called enzymes. Enzymes are *biological catalysts*. A catalyst is a chemical that speeds up a reaction without being used up itself. It takes part in the reaction, but afterwards is unchanged and free to catalyse more reactions. Cells contain hundreds of different enzymes, each catalysing a different reaction. This is how the activities of a cell are controlled – the nucleus contains the genes, which control the production of enzymes, which catalyse reactions in the cytoplasm:

genes → proteins (enzymes) → catalyse reactions

Everything a cell does depends on which enzymes it can make, which in turn depends on which genes in its nucleus are working.

What hasn't been mentioned is why enzymes are needed at all. This is because the temperatures inside organisms are low (e.g. the human body temperature is about 37 °C) and without catalysts, most of the reactions that happen in cells would be far too slow to allow life to go on. Only when enzymes are present to speed them up do the reactions take place quickly enough.

It is possible for there to be thousands of different sorts of enzymes because they are made of proteins, and protein molecules have an enormous range of structures and shapes (see Chapter 4). The molecule that an enzyme acts on is called its **substrate**. Each enzyme has a small area on its surface called the **active site**. The substrate attaches to the active site of the enzyme. The reaction then takes place and products are formed. When the substrate joins up with the active site, it lowers the energy needed for the reaction to start, allowing the products to be formed more easily.

The substrate fits into the active site of the enzyme rather like a key fitting into a lock. That is why this is called the 'lock and key' model of enzyme action (Figure 1.4).

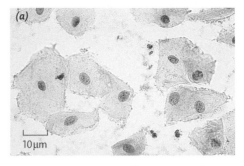

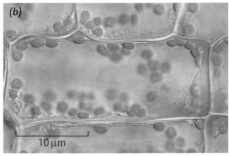

Figure 1.3 *(a) Cells from the lining of a human cheek. (b) Cells from the photosynthetic tissue of a leaf.*

The chemical reactions taking place in a cell are known as **metabolic** reactions. The sum of all the metabolic reactions is known as the **metabolism** of the cell. So the function of enzymes is to catalyse metabolic reactions.

You have probably heard of the enzymes involved in digestion of food. They are secreted by the intestine onto the food to break it down. They are called **extracellular** enzymes, which means 'outside cells'. However, most enzymes stay *inside* cells – they are **intracellular**. You will read about digestive enzymes in Chapter 3.

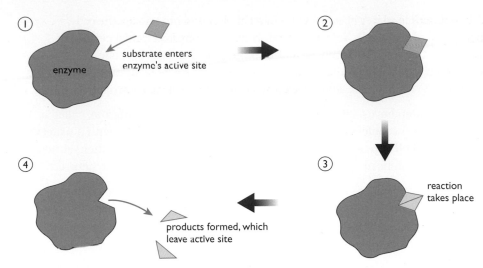

Figure 1.4 *Enzymes catalyse reactions at their active site. This acts like a 'lock' to the substrate 'key'. The substrate fits into the active site, and products are formed. This happens more easily than without the enzyme – so enzymes act as catalysts.*

Notice how, after it has catalysed the reaction once, the enzyme is free to act on more substrate molecules.

Factors affecting enzymes

Temperature affects the action of enzymes. This is easiest to see as a graph, where we plot the rate of the reaction controlled by an enzyme against the temperature (Figure 1.5).

Enzymes in the human body have evolved to work best at about body temperature (37 °C). The graph (Figure 1.5) shows this, because the peak on the curve happens at about this temperature. In this case 37 °C is called the **optimum temperature** for the enzyme.

As the enzyme is heated up to the optimum temperature, increasing temperature speeds up the rate of reaction. This is because higher temperatures give the molecules of enzyme and substrate more energy, so they collide more often. More collisions mean that the reaction will take place more frequently. However, above the optimum temperature another factor comes into play. Enzymes are made of protein, and proteins are broken down by heat. From 40 °C upwards, the heat destroys the enzyme. We say that it is **denatured**. You can see the effect of denaturing when you boil an egg. The egg white is made of protein, and turns from a clear runny liquid into a white solid as the heat denatures the protein.

Temperature is not the only factor that affects an enzyme's activity. The rate of reaction may also be increased by raising the concentration of the enzyme or the substrate. The pH of the surroundings is also important. The pH inside cells is around neutral (pH 7) and not surprisingly, most enzymes have evolved to work best at this pH. At extremes of pH either side of neutral, the enzyme activity decreases, as shown by Figure 1.6. The pH at which the enzyme works best is called the **optimum pH** for that enzyme. Either side of the optimum, the pH affects the structure of the enzyme molecule, and changes the shape of its active site so that the substrate will not fit into it so well.

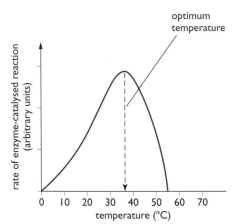

Figure 1.5 *Effect of temperature on the action of an enzyme.*

'Optimum' temperature means the 'best' temperature, in other words the temperature at which the reaction takes place most rapidly.

Not all enzymes have an optimum temperature near 37 °C, just those of animals such as mammals and birds, which all have body temperatures close to this value. Enzymes have evolved to work best at the normal body temperature of the organism. Bacteria that always live at an average temperature of 10 °C will probably have enzymes with an optimum temperature of 10 °C.

Although most enzymes work best at a neutral pH, a few have an optimum below or above pH 7. The stomach produces hydrochloric acid, which makes its contents very acidic (see Chapter 4). Most enzymes stop working at a low pH like this, but the stomach makes an enzyme called pepsin which has an optimum pH of about 2, so that it is adapted to work well in these unusually acidic surroundings.

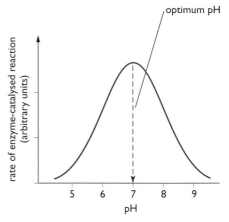

Figure 1.6 *Most enzymes work best at a neutral pH.*

Experiment 1

An investigation into the effect of temperature on the activity of amylase

The digestive enzyme amylase breaks down starch into the sugar maltose. If the speed at which the starch disappears is recorded, this is a measure of the activity of the amylase.

Figure 1.7 shows apparatus which can be used to record how quickly the starch is used up.

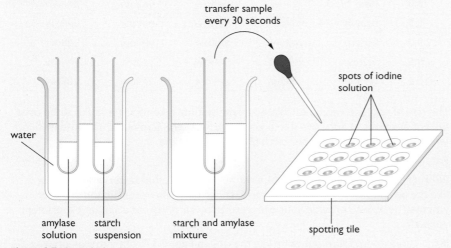

Figure 1.7 *Steps 1–6.*

Spots of iodine are placed into the depressions on the spotting tile. 5 cm^3 of starch suspension is placed in one boiling tube, using a syringe, and 5 cm^3 of amylase solution in another tube, using a different syringe. The beaker is filled with water at room temperature. Both boiling tubes are placed in the beaker of water for 5 minutes, and the temperature recorded.

The amylase solution is then poured into the starch suspension, leaving the tube containing the mixture in the water bath. Immediately, a small sample of the mixture is removed from the tube with a pipette and added to the first drop of iodine on the spotting tile. The colour of the iodine is recorded.

A sample of the mixture is taken every 30 seconds for 10 minutes and tested for starch as above, until the iodine remains yellow, showing that all the starch is used up.

The experiment is repeated, maintaining the water bath at different temperatures between 20 °C and 60 °C. A set of results is shown in the table below.

Time (min)	Colour of mixture at different temperatures				
	20 °C	30 °C	40 °C	50 °C	60 °C
0	blue-black	blue-black	blue-black	blue-black	blue-black
0.5	blue-black	blue-black	brown	blue-black	blue-black
1.0	blue-black	blue-black	yellow	blue-black	blue-black
1.5	blue-black	blue-black	yellow	blue-black	blue-black
2.0	blue-black	blue-black	yellow	brown	blue-black
2.5	blue-black	blue-black	yellow	brown	blue-black
3.0	blue-black	blue-black	yellow	brown	blue-black
3.5	blue-black	blue-black	yellow	yellow	blue-black
4.0	blue-black	blue-black	yellow	yellow	blue-black
5.5	blue-black	blue-black	yellow	yellow	blue-black
6.0	blue-black	brown	yellow	yellow	blue-black
6.5	blue-black	brown	yellow	yellow	blue-black
7.0	blue-black	yellow	yellow	yellow	blue-black
7.5	blue-black	yellow	yellow	yellow	brown
8.0	blue-black	yellow	yellow	yellow	brown
8.5	brown	yellow	yellow	yellow	yellow
9.0	brown	yellow	yellow	yellow	yellow
9.5	yellow	yellow	yellow	yellow	yellow
10.0	yellow	yellow	yellow	yellow	yellow

The rate of reaction can be calculated from the time taken for the starch to be used up. For example, at 50 °C the starch was all gone after 3.5 minutes. The rate is found by dividing the volume of the starch (5 cm³) by the time:

Rate = 5/3.5 = 1.4 cm³/min

Plotting a graph of rate against temperature should produce a curve something like the one shown in Figure 1.6. Try this, either using the results in the table, or you may be able to provide your own results, by carrying out a similar experiment yourself.

If the curve doesn't turn out quite like the one in Figure 1.6, can you explain why this may be? How could you improve the experiment to get more reliable results?

How the cell gets its energy

To be able to carry out all the processes needed for life, a cell needs a source of energy. It gets this by breaking down food molecules to release the stored chemical energy that they contain. This process is called **cell respiration**. Many people think of respiration as meaning 'breathing', but although there are links between the two processes, the biological meaning of respiration is very different.

The process of respiration happens in all the cells of our body. Oxygen is used to oxidise food, and carbon dioxide (and water) are released as waste products. The main food oxidised is glucose (a sugar). Glucose contains stored chemical energy that can be converted into other forms of energy that the cell can use. It is rather like burning a fuel to get the energy out of it, except that burning releases all its energy as heat, whereas respiration releases some heat energy, but most is trapped as energy in other chemicals. This chemical energy can be used for a variety of purposes, such as:

- contraction of muscle cells, producing movement
- active transport of molecules and ions (see page 9)
- building large molecules, such as proteins
- cell division.

The overall reaction for respiration is:

glucose	+	oxygen	$\rightarrow$	carbon dioxide	+	water	(+ energy)
$C_6H_{12}O_6$	+	$6O_2$	$\rightarrow$	$6CO_2$	+	$6H_2O$	(+ energy)

In respiration, carbon passes from glucose out into the atmosphere as carbon dioxide. The carbon can be traced through this pathway using radioactive C^{14}.

This is called **aerobic** respiration, because it uses oxygen. It is not just carried out by human cells, but by all animals and plants and many other organisms. It is important to realise that the equation above is just a *summary* of the process. It actually takes place gradually, as a sequence of small steps that release the energy of the glucose in small amounts. Each step in the process is catalysed by a different enzyme. The later steps in the process are the aerobic ones, and these release the most energy. They happen in the cell's mitochondria.

ATP: the energy 'currency' of the cell

You have seen that respiration gives out energy, while other processes such as protein synthesis and active transport use it up. Cells must have a way of passing the energy from respiration across to these other processes that need it. The way that they do this is through a substance called **adenosine triphosphate**, or **ATP**, which is present in all cells.

ATP is made up from an organic molecule (adenosine) attached to three inorganic phosphate groups (hence triphosphate). ATP can be broken down in the cell, losing a phosphate and producing a similar molecule called adenosine diphosphate, or ADP (Figure 1.8a).

When this reaction happens, energy is released, and is available for the processes that demand energy.

(a) When energy is needed ATP is broken down into ADP and phosphate (P):

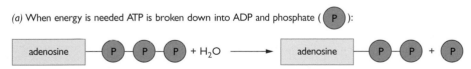

(b) During respiration ATP is made from ADP and phosphate:

Figure 1.8 *ATP is the energy 'currency' of the cell.*

More ATP is made during the reactions of respiration, using the energy from the oxidised glucose to add a phosphate back onto ADP.

Because of its role, ATP is described as the energy 'currency' of a cell. It exchanges chemical energy between the process that produces the energy (respiration) and the processes that use it up.

The reactions of respiration are not 100% efficient, and some energy is not used to make ATP, but instead is lost as heat. Animals such as mammals and birds use this heat to keep their bodies warm, maintaining a constant body temperature (see Chapter 8).

Yeasts can also respire anaerobically. Yeasts are single-celled fungi. When they are deprived of oxygen, they break down sugars into ethanol (alcohol) and carbon dioxide. This is used in commercial processes such as making wine and beer, and baking bread.

There are some situations where cells can respire *without* using oxygen. This is called **anaerobic** respiration. In anaerobic respiration, glucose is not completely broken down, and less energy is released. However, the advantage of anaerobic respiration is that it can occur in situations where oxygen is in short supply, for example in contracting muscle cells.

If muscles are overworked, the blood cannot reach them fast enough to deliver enough oxygen for aerobic respiration. This happens when a person does a 'burst' activity, such as a sprint, or quickly lifting a heavy weight. The glucose is broken down into a substance called **lactic acid**:

$$\text{glucose} \quad \rightarrow \quad \text{lactic acid} \qquad (+ \text{ some energy})$$
$$C_6H_{12}O_6 \quad \rightarrow \quad 2C_3H_6O_3$$

It used to be thought that lactic acid was a cause of cramps in overused muscles. It is now known that this is not true. Muscle cramp may happen, but it is caused by various other factors resulting from the intense exercise, and is not due to the lactic acid.

Anaerobic respiration provides enough energy to keep the overworked muscles going for a short period, but continuing the 'burst' activity makes lactic acid build up in the bloodstream. When the period of exercise is over, the lactic acid is oxidised aerobically. This uses oxygen. It takes about 30 minutes of rest for all the lactic acid to be used up (see Chapter 4, page 66). The volume of oxygen needed to completely oxidise the lactic acid that builds up in the body during anaerobic respiration is called the **oxygen debt**.

Movements of materials in and out of cells

Cell respiration shows the need for cells to be able to take in certain substances from their surroundings, such as glucose and oxygen, and get rid of others, such as carbon dioxide and water. As you have seen, the cell surface membrane is selective about which chemicals can pass in and out. There are three main ways that molecules and ions can move through the membrane. They are diffusion, active transport and osmosis.

Diffusion is the net movement of particles (molecules or ions) from a region of high concentration to a region of low concentration, i.e. down a concentration gradient.

Many substances can pass through the membrane by **diffusion**. Diffusion happens when a substance is more concentrated in one place than another. For example, if the cell is making carbon dioxide by respiration, the concentration of carbon dioxide inside the cell will be higher than outside. This difference in concentration is called a **concentration gradient**. The molecules of carbon dioxide are constantly moving about because of their kinetic energy. The cell membrane is permeable to carbon dioxide, so they can move in either direction through it.

Because there is a higher concentration of carbon dioxide molecules inside the cell than outside, over time more molecules will move from inside the cell to outside than move in the other direction. We say that there is a *net* movement of the molecules from inside to outside (Figure 1.9).

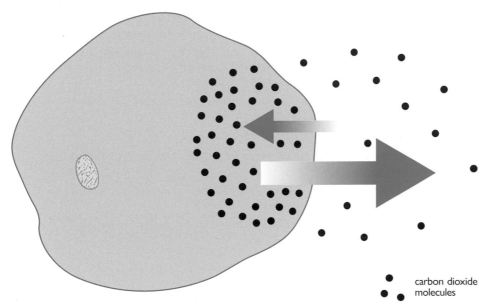

carbon dioxide
molecules

Figure 1.9 *Carbon dioxide is produced by respiration, so its concentration builds up inside the cell. Although the carbon dioxide molecules diffuse in both directions across the cell membrane, the overall (net) movement is out of the cell, down the concentration gradient.*

The opposite happens with oxygen. Respiration uses up oxygen, so there is a concentration gradient of oxygen from outside to inside the cell. There is therefore a net movement of oxygen *into* the cell by diffusion.

Diffusion happens because of the kinetic energy of the particles. It does not need an 'extra' source of energy from respiration. However, sometimes a cell needs to take in a substance when there is very little of that substance outside the cell, in other words *against* a concentration gradient. It can do this by another process, called **active transport**. The cell uses energy from respiration to take up the particles, rather like a pump uses energy to move a liquid from one place to another. In fact, biologists usually speak of the cell 'pumping' ions or molecules in or out. The pumps are large protein molecules located in the cell membrane. An example of a place where this happens is in the human small intestine, where some glucose in the gut is absorbed into the cells lining the intestine by active transport.

The rate of diffusion of a substance is greater at higher temperatures. The reason for this is that a higher temperature will give the diffusing particles more kinetic energy.

The rate of diffusion of a substance is increased by:
• a steep concentration gradient
• high temperatures
• a large surface area to volume ratio.

Active transport is the movement of particles against a concentration gradient, using energy from respiration.

Demonstration of diffusion in a jelly

Agar jelly has a consistency similar to the cytoplasm of a cell. Like cytoplasm, it has a high water content. Agar can be used to show how substances diffuse through a cell.

This demonstration uses the reaction between hydrochloric acid and potassium permanganate solution. When hydrochloric acid comes into contact with potassium permanganate, the purple colour of the permanganate disappears.

A Petri dish is prepared which contains a 2 cm deep layer of agar jelly, dyed purple with potassium permanganate. Three cubes of different sizes are cut out of the jelly, with side lengths 2 cm, 1 cm and 0.5 cm.

The cubes are carefully dropped, at the same time, into a beaker of dilute hydrochloric acid (Figure 1.10)

The time is taken for each cube to turn colourless.

Which cube would be the first to turn colourless and which the last? Explain the reasoning behind your prediction.

If the three cubes represented cells of different sizes, which cell would have the most difficulty in obtaining substances by diffusion?

It may be possible for you to try this experiment, using similar apparatus.

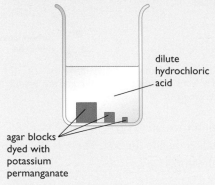

dilute hydrochloric acid

agar blocks dyed with potassium permanganate

Figure 1.10 *Investigating diffusion in a jelly.*

Earlier in this chapter we called the cell membrane 'selectively' permeable. This term is sometimes used when describing osmosis. It means that the membrane has control over which molecules it lets through (e.g. by active transport). 'Partially' permeable just means that small molecules such as water and gases can pass through, while larger molecules cannot. Strictly, the two words are not interchangeable, but they are often used this way in biology books.

Osmosis in cells is the net movement of water from a dilute solution to a more concentrated solution across the partially permeable cell membrane.

Water moves across cell membranes by a special sort of diffusion, called **osmosis**. Osmosis happens when the total concentrations of all dissolved substances inside and outside the cell are different. Water will move across the membrane from the more dilute solution to the more concentrated one. Notice that this is still obeying the rules of diffusion – the water moves from where there is a higher concentration of *water* molecules to a lower concentration of *water* molecules. Osmosis can only happen if the membrane is permeable to water but not to some other solutes. We say that it is **partially** permeable.

One artificial partially permeable membrane is called Visking tubing. This is used in kidney dialysis machines. Visking tubing has microscopic holes in it, which let small molecules like water pass through (it is *permeable* to them) but is not permeable to some larger molecules, such as the sugar sucrose. This is why it is called 'partially' permeable. You can show the effects of osmosis by filling a Visking tubing 'sausage' with concentrated sucrose solution, attaching it to a capillary tube and placing the Visking tubing in a beaker of water (Figure 1.11).

The level in the capillary tube rises as water moves from the beaker to the inside of the Visking tubing. This movement is due to osmosis. You can understand what's happening if you imagine a highly magnified view of the Visking tubing separating the two liquids (Figure 1.12).

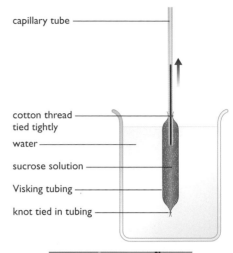

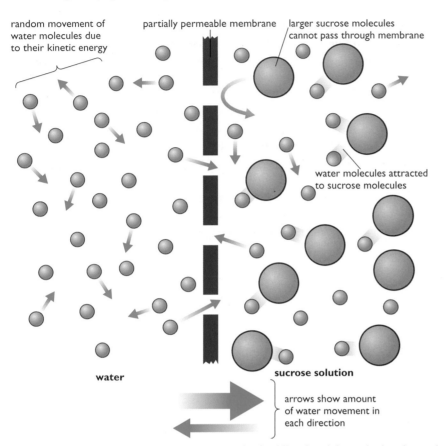

Figure 1.12 *In this model of osmosis, more water molecules diffuse from left to right than from right to left.*

Figure 1.11 *Water enters the Visking tubing 'sausage' by osmosis. This causes the level of liquid in the capillary tube to rise. In the photograph, the contents of the Visking tubing have had a red dye added to make it easier to see the movement of the liquid.*

The sucrose molecules are too big to pass through the holes in the partially permeable membrane. The water molecules can pass through the membrane in either direction, but those on the right are attracted to the sugar molecules. This slows them down and means that they are less free to move – they have less kinetic energy. As a result of this, more water molecules diffuse from left to right than from right to left. In other words, there is a greater diffusion of water molecules from the more dilute solution (in this case pure water) to the more concentrated solution.

How 'free' the water molecules are to move is called the **water potential**. The molecules in pure water can move most freely, so pure water has the highest water potential. The more concentrated a solution is, the lower is its water potential. In the model in Figure 1.12, water moves from a high to a low water potential. This is a law that applies whenever water moves by osmosis. We can bring these ideas together in an alternative definition of osmosis.

It is important to realise that neither of the two solutions has to be pure water. As long as there is a difference in their concentrations (and their water potentials), and they are separated by a partially permeable membrane, osmosis can still take place.

Osmosis is the net diffusion of water across a partially permeable membrane, from a solution with a high water potential to one with a lower water potential.

All cells are surrounded by a partially permeable cell membrane. In the human body, osmosis is important in moving water from cell to cell, and from the blood to the tissues (see Chapter 4, page 60). It is important that the cells of the body are bathed in a solution with the right concentration of solutes; otherwise they could be damaged by osmotic movements of water. For example, if red blood cells are put into water, they will swell up and burst. If the same cells are put into a concentrated salt solution, they lose water by osmosis and shrink, producing cells with crinkly edges.

Experiment 3

Demonstration of the effects of osmosis on red blood cells

Blood plasma has a concentration equivalent to a 0.85% salt solution. If fresh blood is placed into solutions with different concentrations, the blood cells will gain or lose water by osmosis. This can be demonstrated using sterile animal blood (available from suppliers of biological materials).

Three test tubes are set up, containing these solutions:

A 10 cm³ of distilled water

B 10 cm³ of 0.85% salt solution

C 10 cm³ of 3% salt solution

1 cm³ of blood is added to each tube, and the tubes are shaken. A sample from each tube is examined under the microscope. The sample from tube A is found to contain no intact cells. Figure 1.13 shows cells from tubes B and C. The cells from tube B look normal, but those from tube C are shrunken, with crinkly edges.

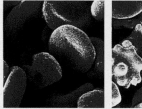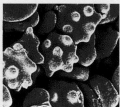

Figure 1.13 *Compare the blood cells on the right, which were placed in a 3% salt solution, with the normal blood cells on the left, from a 0.85% salt solution.*

Using your knowledge of osmosis, can you explain what happens to the red blood cells in each tube?

The three tubes are now placed in a centrifuge and spun around at high speed to separate any solid particles from solution. The results are shown in Figure 1.14.

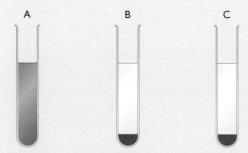

Figure 1.14 *The three tubes after centrifugation.*

Tube A contains a clear red solution and no solid material at the bottom of the tube. Tubes B and C both contain a colourless liquid and a red precipitate at the bottom.

Can you explain these results?

Experiment 3 shows how important it is that animal cells are surrounded by a solution containing the correct concentration of dissolved solutes. If the surrounding solution does not have the right concentration, cells can be damaged by the effects of osmosis. The red blood cells placed in water absorb the water by osmosis, swell up and burst, leaving a red solution of haemoglobin in the test tube. When placed in 3% salt solution, the red blood cells lose water by osmosis and shrink.

We will return to the idea that cells need a correct constant 'environment' in Chapter 8.

All cells exchange substances with their surroundings, but some parts of the body are specially adapted for the exchange of materials because they have a very large surface area in proportion to their volume. Two examples are the alveoli of the lungs (Chapter 2) and the villi of the small intestine (Chapter 3). Diffusion is a slow process, and organs that rely on diffusion need a large surface over which it can take place. The alveoli (air sacs) allow exchange of oxygen and carbon dioxide to take place between the air and the blood, during breathing. The villi of the small intestine provide a large surface area for the absorption of digested food.

Cell division and differentiation

Humans begin life as a single fertilised egg cell, called a **zygote**. This divides into two cells, then four, then eight and so on, until the adult body contains countless millions of cells (Figure 1.15).

This type of cell division is called **mitosis** and is under the control of the genes. You can read a full account of mitosis in Chapter 11, but it is worthwhile considering an outline of the process now. First of all the chromosomes in the nucleus are copied, then the nucleus splits into two, so that the genetic information is shared equally between the two 'daughter' cells. The cytoplasm then divides, forming two smaller cells. These then take in food substances to supply energy and building materials so that they can grow to full size. The process is repeated, but as the developing **embryo** grows, cells become specialised to carry out particular roles. This specialisation is also under the control of the genes, and is called **differentiation**. Different kinds of cells develop depending on where they are located in the embryo, for example a nerve cell in the spinal cord, or an epidermal cell in the outer layer of the skin (Figure 1.16). Throughout this book you will read about cells that have a structure adapted for a particular function.

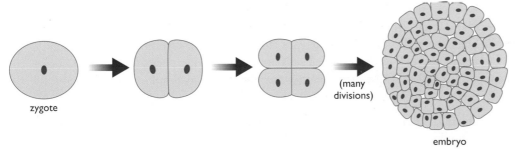

zygote

(many divisions)

embryo

Figure 1.15 *Animals and plants grow by cell division.*

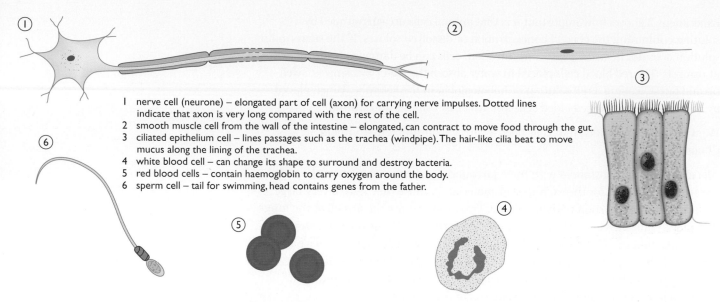

1. nerve cell (neurone) – elongated part of cell (axon) for carrying nerve impulses. Dotted lines indicate that axon is very long compared with the rest of the cell.
2. smooth muscle cell from the wall of the intestine – elongated, can contract to move food through the gut.
3. ciliated epithelium cell – lines passages such as the trachea (windpipe). The hair-like cilia beat to move mucus along the lining of the trachea.
4. white blood cell – can change its shape to surround and destroy bacteria.
5. red blood cells – contain haemoglobin to carry oxygen around the body.
6. sperm cell – tail for swimming, head contains genes from the father.

Figure 1.16 *Some cells with very specialised functions. They are not drawn to the same scale.*

What is hard to understand about this process is that through mitosis all the cells of the body have the *same* genes. How is it that some genes are 'switched on' and others are 'switched off' to produce different cells? The answer to this question is very complicated, and scientists are only just beginning to work it out.

Cells, tissues and organs

Cells that have a similar function are grouped together as **tissues**. For example the muscle of your arm contains millions of muscle cells, all specialised for one function – contraction to move the arm bones (Chapter 7). This is muscle tissue, or more accurately **voluntary muscle** tissue. The 'voluntary' refers to the fact that contraction of muscles like this is under conscious control of the brain. The smooth muscle cell shown in Figure 1.16 above makes up **involuntary muscle**, since the gut muscles are not under voluntary control by your brain. Involuntary muscle is present in the walls of organs such as the intestine, bladder and blood vessels. There is a third type of muscle tissue called **cardiac** muscle, which makes up the muscular wall of the heart.

Tissues that line organs are called **epithelia** (singular 'epithelium'). Figure 1.16 shows a **ciliated** epithelium cell, which has minute hair-like projections called cilia, able to beat to move materials along. There are several other types of epithelia, such as the flattened cells lining the human cheek (Figure 1.3). This is called a **squamous** epithelium. You will read about several types of epithelia throughout this book.

Bone is a tissue made of cells that secrete a hard matrix made of calcium salts (Chapter 7). Other tissues include **blood** (Chapter 4) which is made of various types of red and white blood cells in a liquid matrix called plasma, and **nervous tissue** (Chapter 5), which makes up the brain, spinal cord and nerves.

A collection of different tissues carrying out a particular function is called an **organ**. The main organs of the human body are shown in Figure 1.17.

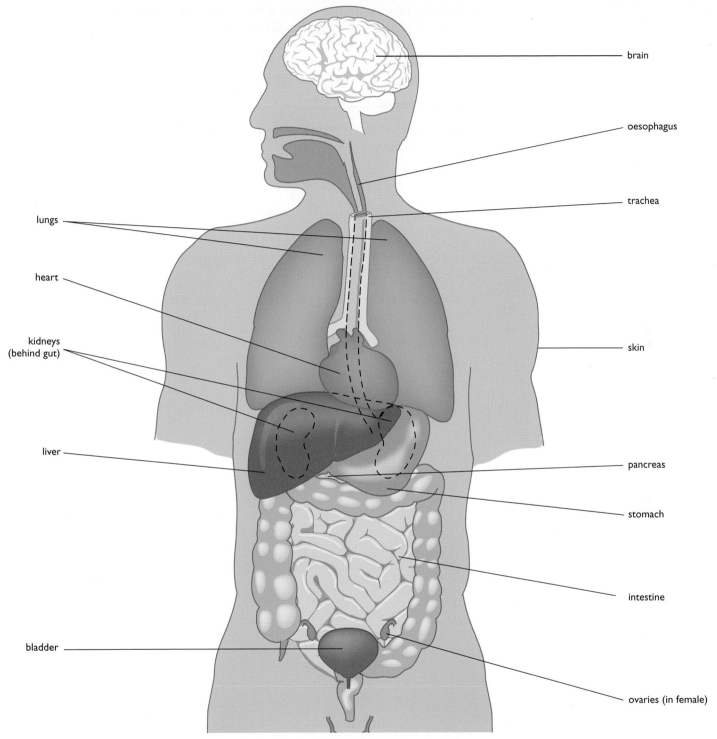

brain

oesophagus

trachea

lungs

heart

kidneys
(behind gut)

skin

liver

pancreas

stomach

intestine

bladder

ovaries (in female)

Figure 1.17 *Some of the main organs of the human body.*

In humans, jobs are usually carried out by several different organs working together. This is called an **organ system**. For example, the digestive system consists of the gut, along with glands such as the pancreas and gall bladder. The function of the whole system is to digest food and absorb the digested products into the blood. There are seven main systems in the human body, these are the:

- **digestive** system

- **respiratory** system – including the lungs, which exchange oxygen and carbon dioxide

- **circulatory** system – including the heart and blood vessels, which transport materials around the body

- **excretory** system – including the kidneys, which filter toxic waste materials from the blood

- **nervous** system – consisting of the brain, spinal cord and nerves, which coordinate the body's actions

- **endocrine** system – glands secreting hormones, which act as chemical messengers

- **reproductive** system – producing sperm in males and eggs in females, and allowing the development of the embryo.

You should now be able to:

✓ recognise cell structures and describe the functions of the nucleus, cytoplasm, cell membrane, mitochondria, endoplasmic reticulum and ribosomes

✓ explain the role and functioning of enzymes as biological catalysts in metabolic reactions

✓ explain how enzymes can be affected by changes in temperature and pH

✓ describe how to carry out simple controlled experiments to show how enzyme activity can be affected by changes in temperature

✓ understand the process of aerobic respiration and recall the word equation and chemical symbol equation for aerobic respiration

✓ explain the significance of the breakdown and regeneration of ATP

✓ explain the differences between aerobic and anaerobic respiration and the formation of lactic acid in anaerobic respiration

✓ understand the movement of substances into and out of cells by diffusion, osmosis and active transport, and the factors that can affect the rate of movement

✓ describe how to carry out simple experiments on diffusion and osmosis using living and non-living systems

✓ understand the grouping of cells into tissues, including: voluntary, involuntary and cardiac muscle; bone, blood, nervous tissue and squamous and ciliated epithelia

✓ recall the organisation of cells into organs.

Questions

1 a) Draw a diagram of an animal cell. Label all of the parts. Alongside each label write the function of that part.

 b) Write down three differences between the cell you have drawn and a 'typical' plant cell.

2 Write a short description of the nature and function of enzymes. It would be easier if you worked on a computer. Include in your description:

- a definition of an enzyme

- a description of the 'lock and key' model of enzyme action

- an explanation of the difference between intracellular and extracellular enzymes.

Your description should be about a page in length, including a labelled diagram.

3 The graph shows the effect of temperature on an enzyme. The enzyme was extracted from a microorganism that lives in hot mineral springs near a volcano.

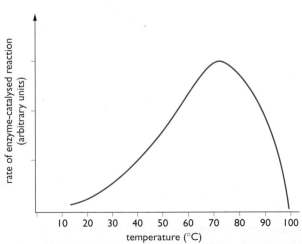

a) What is the optimum temperature of this enzyme?

b) Explain why the activity of the enzyme is greater at 60 °C than at 30 °C.

c) The optimum temperature of enzymes in the human body is about 37 °C. Explain why this enzyme is different.

d) What happens to the enzyme at 90 °C?

4 Explain the differences between diffusion and active transport.

5 The nerve cell called a **motor neurone** (page 14) and a **red blood cell** (page 14) are both very specialised cells. Look up each of these cells in this book and explain very briefly (three or four lines) how each is adapted to its function.

6 The diagram shows a cell from the lining of a human kidney tubule. A major role of the cell is to absorb glucose from the fluid passing along the tubule and pass it into the blood, as shown by the arrows on the diagram.

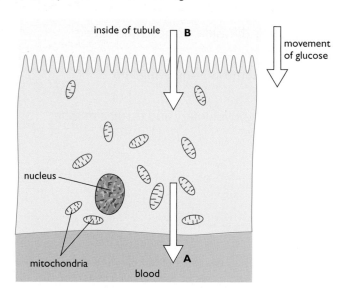

a) What is the function of the mitochondria?

b) The tubule cell contains a large number of mitochondria. They are needed for the cell to transport glucose across the cell membrane into the blood at 'A'. Suggest the method that the cell uses to do this and explain your answer.

c) The mitochondria are not needed to transport the glucose into the cell from the tubule at 'B'. Name the process by which the ions move across the membrane at 'B' and explain your answer.

d) The surface membrane of the tubule cell at 'B' is greatly folded. Explain how this adaptation helps the cell to carry out its function.

7 An experiment was carried out to find the effects of osmosis on blood cells. Three test tubes were filled with different solutions. 10 cm³ of water was placed in tube A, 10 cm³ of 0.85% salt solution in tube B, and 10 cm³ of 3% salt solution in tube C. 1 cm³ of fresh blood was added to each tube. The tubes were shaken, and then a sample from each was observed under the microscope under high power.

The tubes were then placed in a centrifuge and spun around at high speed to separate any solid particles from solution. The results are shown in the diagram below.

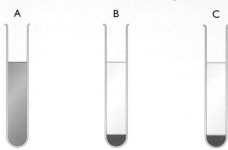

a) Which solution had a similar salt concentration to blood?

b) Describe what you would expect to see when viewing the samples from tubes A to C through the microscope.

c) Explain the results shown in the diagram.

d) When a patient has suffered severe burns, damage to the skin results in a loss of water from the body. This condition can be treated by giving the patient a saline drip. This is a 0.85% salt solution which is fed into the patient's blood through a needle inserted into a vein. Explain why 0.85% salt solution is used, and not water.

8 In multicellular organisms, cells are organised into tissues, organs and organ systems.

a) The diagram shows a section through an artery and a capillary.

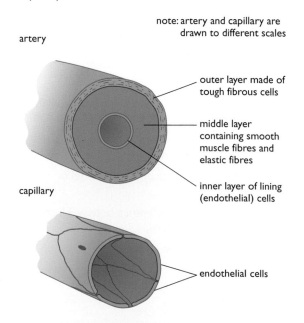

Explain why an artery can be considered to be an organ whereas a capillary cannot.

b) Organ systems contain two or more organs whose functions are linked. The digestive system is one human organ system. (See Chapter 3.)

   i) What does the digestive system do?

   ii) Name three organs in the human digestive system. Explain what each organ does as part of the digestive system.

   iii) Name two other human organ systems and, for each system, name two organs that are part of the system.

9 Different particles move across cell membranes using different processes. The table below shows some ways in which active transport, osmosis and diffusion are similar and some ways in which they are different. Copy and complete the table with ticks and crosses.

Feature	Active transport	Osmosis	Diffusion
particles must have kinetic energy			
requires energy from respiration			
particles move down a concentration gradient			
process needs special carriers in the membrane			

10 Cells in the wall of the small intestine divide by mitosis to replace cells lost as food passes through.

a) Chromosomes contain DNA. The graph shows the changes in the DNA content of a cell in the wall of the small intestine as it divides by mitosis.

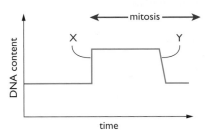

   i) Why is it essential that the DNA content is doubled (X) before mitosis commences?

   ii) What do you think happens to the cell at point Y?

b) The diagram shows a cell in the wall of a villus in the small intestine (see Chapter 3, page 47). The cell absorbs glucose from the intestine and passes it into the blood.

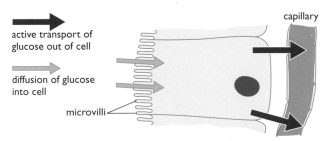

   i) Suggest how the microvilli adapt this cell to its function of absorbing glucose.

   ii) Suggest how the active transport of glucose out of the cell and into the bloodstream helps with the absorption of glucose from the small intestine.

Chapter 2: Breathing and Gas Exchange

When we breathe, air is drawn into and out of the lungs so that gas exchange can take place between the air and the blood. This chapter looks at these processes, and also deals with some ways that smoking can damage the lungs and stop these vital organs from working properly.

Cells get their energy by oxidising foods such as glucose. This process is called cell respiration (see Chapter 1). If the cells are to respire aerobically, they need a continuous supply of oxygen from the blood. In addition, the waste product of respiration, carbon dioxide, needs to be removed from the body. In humans, these gases are exchanged between the blood and the air in the lungs.

Respiration and breathing

We need to understand the difference between respiration and breathing. Respiration is the oxidation reaction (described in Chapter 1) that releases energy from foods, such as glucose. We can use the term **ventilation** to describe the mechanism that moves air into and out of the lungs, which then allows gas exchange to take place. We also use the term **breathing** in a more general way for this process. The lungs and associated structures are often called the 'respiratory system' but this can be confusing. It is better to call them the **gas exchange system** and this is the term we adopt in this book.

The structure of the gas exchange system

The lungs are enclosed in the chest or **thorax** by the ribcage and a muscular sheet of tissue called the **diaphragm** (Figure 2.1). As you will see, the actions of these two structures bring about the movements of air into and out of the lungs. Joining

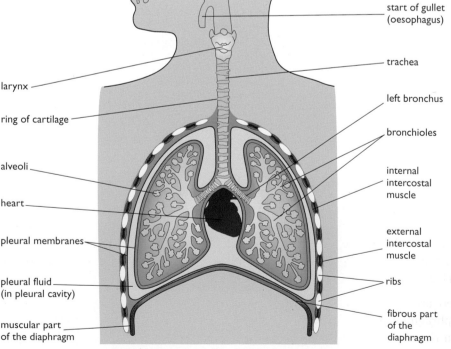

Figure 2.1 *The human gas exchange system.*

each rib to the next are two sets of muscles called **intercostal muscles** ('costals' are rib bones). If you eat meat, you will have seen intercostal muscle attached to the long bones of 'spare ribs'. The diaphragm separates the contents of the thorax from the abdomen. It is not flat, but a shallow dome shape, with a fibrous middle part forming the 'roof' of the dome, and muscular edges forming the walls.

The air passages of the lungs form a highly branching network (Figure 2.2). This is why it is sometimes called the **bronchial tree**.

When we breathe in, air enters our nose or mouth and passes down the windpipe or **trachea**. The trachea splits into two tubes called the **bronchi**, one leading to each lung. Each **bronchus** divides into smaller and smaller tubes called **bronchioles**, eventually ending at microscopic air sacs, called **alveoli**. It is here that gas exchange with the blood takes place.

The walls of trachea and bronchi contain rings of gristle or **cartilage**. These support the airways and keep them open when we breathe in. They are rather like the rings in a vacuum cleaner hose – without them the hose would squash flat when the cleaner sucks air in.

The inside of the thorax is separated from the lungs by two thin, moist membranes called the **pleural membranes**. They make up a continuous envelope around the lungs, forming an airtight seal. Between the two membranes is a space called the **pleural cavity**, filled with a thin layer of liquid called **pleural fluid**. This acts as lubrication, so that the surfaces of the lungs don't stick to the inside of the chest wall when we breathe.

Keeping the airways clean

The trachea and larger airways are lined with a layer of cells that have an important role in keeping the airways clean. Some cells in this lining secrete a sticky liquid called **mucus**, which traps particles of dirt or bacteria that are breathed in. Other cells are covered with tiny hair-like structures called **cilia** (Figure 2.4). The cilia beat backwards and forwards, sweeping the mucus and trapped particles out towards the mouth. In this way, dirt and bacteria are prevented from entering the lungs, where they might cause an infection. As you will see, one of the effects of smoking is that it destroys the cilia and stops this protection mechanism from working properly.

Ventilation of the lungs

Ventilation means moving air in and out of the lungs. This requires a difference in air pressure – the air moves from a place where the pressure is high to one where it is low. Ventilation depends on the fact that the thorax is an airtight cavity. When we breathe, we change the volume of our thorax, which alters the pressure inside it. This causes air to move in or out of the lungs.

There are two movements that bring about ventilation, those of the ribs and the diaphragm. If you put your hands on your chest and breathe in deeply, you can feel your ribs move upwards and outwards. They are moved by the intercostal muscles (Figure 2.5). The outer (external) intercostals contract, pulling the ribs up. At the same time, the muscles of the diaphragm contract, pulling the diaphragm down into a more flattened shape (Figure 2.6a). Both these movements increase the volume of the chest and cause a slight drop in pressure inside the thorax compared with the air outside. Air then enters the lungs.

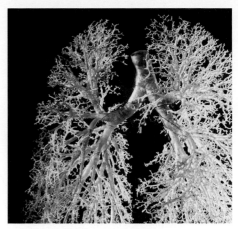

Figure 2.2 *This cast of the human lungs was made by injecting a pair of lungs with a liquid plastic. The plastic was allowed to set, then the lung tissue was dissolved away with acid.*

In the bronchi, the cartilage forms complete, circular rings. In the trachea, the rings are incomplete, and shaped like a letter 'C'. The open part of the ring is at the back of the trachea, next to where the oesophagus (gullet) lies as it passes through the thorax. When food passes along the oesophagus by peristalsis (see Chapter 3) the gaps in the rings allow the lumps of food to pass through more easily, without the peristaltic wave 'catching' on the rings (Figure 2.3).

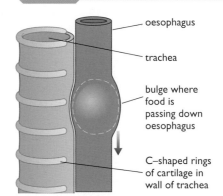

oesophagus

trachea

bulge where food is passing down oesophagus

C–shaped rings of cartilage in wall of trachea

Figure 2.3 *C-shaped cartilage rings in the trachea.*

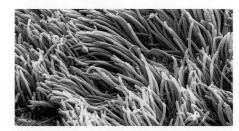

Figure 2.4 *This electron microscope picture shows cilia from the lining of the trachea.*

During normal (shallow) breathing, the elasticity of the lungs and the weight of the ribs acting downwards are enough to cause exhalation. The internal intercostals are only really used for deep (forced) breathing out, for instance when we are exercising.

It is important that you remember the changes in volume and pressure during ventilation. If you have trouble understanding these, think of what happens when you use a bicycle pump. If you push the pump handle, the air in the pump is squashed, its pressure rises and it is forced out of the pump. If you pull on the handle, the air pressure inside the pump falls a little, and air is drawn in from outside. This is similar to what happens in the lungs. In exams, students sometimes talk about the lungs *forcing* the air in and out – they don't!

The opposite happens when you breathe out deeply. The external intercostals relax, and the internal intercostals contract, pulling the ribs down and in. At the same time, the diaphragm muscles relax and the diaphragm goes back to its normal dome shape. The volume of the thorax decreases, and the pressure in the thorax is raised slightly above atmospheric pressure. This time the difference in pressure forces air out of the lungs (Figure 2.6b). Exhalation is helped by the fact that the lungs are elastic, so that they tend to empty like a balloon.

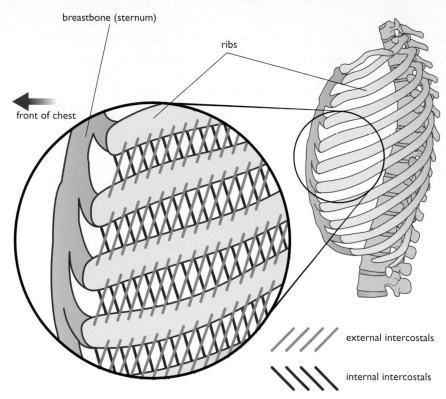

Figure 2.5 *Side view of the chest wall, showing the ribs. The diagram shows how the two sets of intercostal muscles run between the ribs. When the external intercostals contract, they move the ribs upwards. When the internal intercostals contract, the ribs are moved downwards.*

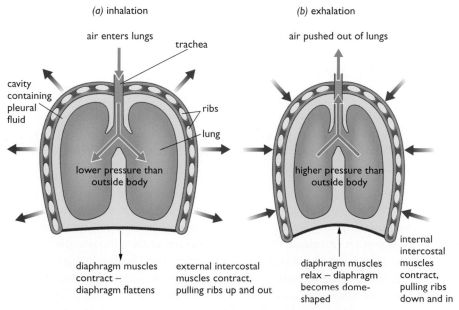

Figure 2.6 *Changes in the position of the ribs and diaphragm during breathing. (a) Breathing in (inhalation). (b) Breathing out (exhalation).*

Gas exchange in the alveoli

You can tell what is happening during gas exchange if you compare the amounts of different gases in atmospheric air with the air breathed out (Table 2.1).

Gas	Atmospheric air	Exhaled air
nitrogen	78	79
oxygen	21	16
carbon dioxide	0.04	4
other gases (mainly argon)	1	1

Table 2.1: *Approximate percentage volume of gases in atmospheric (inhaled) and exhaled air.*

Exhaled air is also warmer than atmospheric air, and is saturated with water vapour. The amount of water vapour in the atmosphere varies, depending on weather conditions.

Clearly, the lungs are absorbing oxygen into the blood and removing carbon dioxide from it. This happens in the alveoli. To do this efficiently, the alveoli must have a structure which brings the air and blood very close together, over a very large surface area. There are enormous numbers of alveoli. It has been calculated that the two lungs contain about 700 000 000 of these tiny air sacs, giving a total surface area of 60 m². That's bigger than the floor area of an average classroom! Viewed through a high-powered microscope, the alveoli look rather like bunches of grapes, and are covered with tiny blood capillaries (Figure 2.7).

Be careful when interpreting percentages! The *percentage* of a gas in a mixture can vary, even if the actual *amount* of the gas stays the same. This is easiest to understand from an example. Imagine you have a bottle containing a mixture of 20% oxygen and 80% nitrogen. If you used a chemical to absorb all the oxygen in the bottle, the nitrogen left would now be 100% of the gas in the bottle, despite the fact that the *amount* of nitrogen would still be the same. That is why the percentage of nitrogen in inhaled and exhaled air is slightly different.

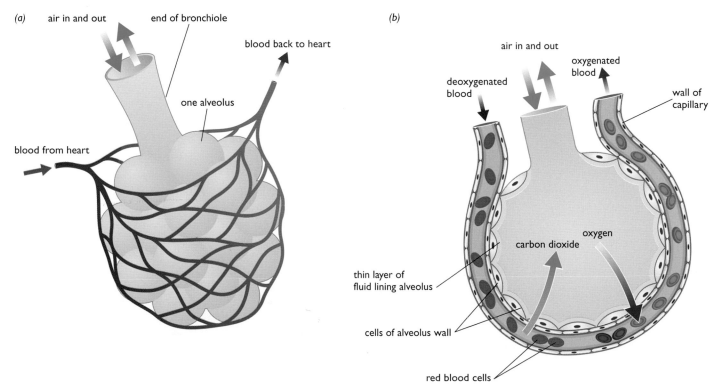

Figure 2.7 *(a) Alveoli and the surrounding capillary network. (b) Diffusion of oxygen and carbon dioxide takes place between the air in the alveolus and the blood in the capillaries.*

The thin layer of fluid lining the inside of the alveoli comes from the blood. The capillaries and cells of the alveolar wall are 'leaky' and the blood pressure pushes fluid out from the blood plasma into the alveolus. Oxygen dissolves in this moist surface before it passes through the alveolar wall into the blood.

Blood is pumped from the heart to the lungs and passes through the capillaries surrounding the alveoli. The blood has come from the respiring tissues of the body, where it has given up some of its oxygen to the cells, and gained carbon dioxide. Around the lungs, the blood is separated from the air inside each alveolus by only two cell layers; the cells making up the wall of the alveolus, and the capillary wall itself. This is a distance of less than a thousandth of a millimetre.

Because the air in the alveolus has a higher concentration of oxygen than the blood entering the capillary network, oxygen diffuses from the air, across the wall of the alveolus and into the blood. At the same time there is more carbon dioxide in the blood than there is in the air in the lungs. This means that there is a diffusion gradient for carbon dioxide in the other direction, so carbon dioxide diffuses the other way, out of the blood and into the alveolus. The result is that the blood which leaves the capillaries and flows back to the heart has gained oxygen and lost carbon dioxide. The heart then pumps the blood around the body again, to supply the respiring cells (see Chapter 4).

Experiment 4

Comparing the carbon dioxide content of inhaled and exhaled air

The apparatus in Figure 2.8 can be used to compare the amount of carbon dioxide in inhaled and exhaled air. A person breathes gently in and out through the middle tube. Exhaled air passes out through one tube of indicator solution and inhaled air is drawn in through the other tube. If limewater is used, the limewater in the 'exhaled' tube will turn cloudy before the limewater in the 'inhaled' tube. (If hydrogencarbonate indicator solution is used instead, it changes from red to yellow.)

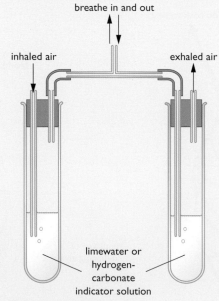

Figure 2.8 *Apparatus for Experiment 4.*

An investigation into the effect of exercise on breathing rate

It is easy to show the effect of exercise on breathing rate in an experimental 'subject'. He (or she) sits quietly for 5 minutes, making sure that he is completely relaxed. He then counts the number of breaths he takes in 1 minute, recording his results in a table. He waits a minute, and then counts his breaths again, recording the result, and repeating if necessary until he gets a steady value for the 'resting rate'.

The subject then carries out some vigorous exercise, such as running on a treadmill for 3 minutes. Immediately he finishes the exercise, he sits down and records the breathing rate as before. He then continues to record his breaths per minute, every minute, until he returns to his normal resting rate.

The table shows the results from an investigation into the breathing rate of two girls, A and B, before and after exercise.

Plot a line graph of these results, using the same axes for both subjects. Join the data points using straight lines, and leave a gap during the period of exercise, when no readings were taken.

Why does breathing rate need to rise during exercise? Explain as fully as possible. Why does the rate not return to normal as soon as a subject finishes the exercise? (See Chapter 1.)

Describe the difference in the breathing rates of the two girls (A and B) after exercise. Which girl is more fit? Explain your reasoning.

Time from start of expt (min)	Breathing rate (breaths/min)	
	A	B
1	13	13
2	14	12
3	14	12
Rate after 3 min vigorous exercise:		
7	28	17
8	24	13
9	17	12
10	14	12

Measuring breathing volumes

Breathing volumes can be measured using a piece of apparatus called a **spirometer** (Figure 2.9).

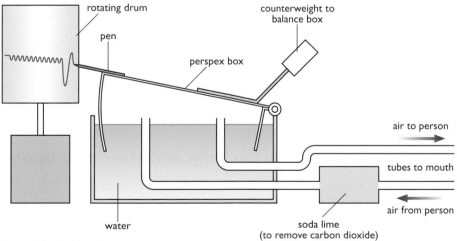

Figure 2.9 *A spirometer.*

The floating box is filled with air, or with medical-grade oxygen. A person (the 'subject' of the experiment) breathes in and out through a mouthpiece into the floating lid of the spirometer chamber. As they breathe, the lid moves up and down, and a pen records the breathing movements on a rotating drum, which is covered by a piece of graph paper. A canister of soda lime absorbs carbon dioxide from the subject's exhaled air. The speed at which the drum turns is set, so that the number of breaths per minute can be worked out.

Figure 2.10 shows a trace from a subject breathing into a spirometer. The subject began breathing normally into the spirometer, followed by a deep breath in. He then breathed out as much as possible.

The volume of air breathed in and out with a normal breath is called the tidal volume. In a normal healthy adult man, this is about $0.4\,dm^3$. The difference between the maximum breath in and maximum breath out is about $4.5\,dm^3$. This is called the **vital capacity**.

The spirometer trace does not show the full amount of air in the lungs. There is always about $1.5\,dm^3$ of air left in the lungs, even after a person has breathed out as much as possible. This is called the **residual volume**.

> The breathing volumes depend on various factors, including age, sex of the person, size of the lungs, level of exercise and how healthy the subject is. The vital capacity varies from about $3.0\,dm^3$ in a young woman to $6.0\,dm^3$ in a trained athlete.

> Notice that the trace in Figure 2.10 is not horizontal, but slowly slopes down to the right. This is because the oxygen in the spirometer is gradually being used up by the person's respiration, and is not replaced by carbon dioxide.

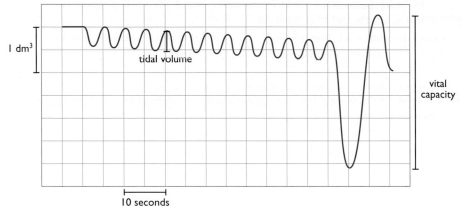

Figure 2.10 *A spirometer trace showing breathing volumes.*

Control of breathing rate

When you exercise, your muscles need more oxygen for respiration, and they produce more carbon dioxide, which has to be removed from the body. To bring about these changes your breathing rate increases. This happens automatically. As the carbon dioxide builds up in the blood, it is detected by receptors in the brain and other locations. These receptors send nerve impulses to the diaphragm and intercostal muscles, causing both the breathing rate and depth of breathing to increase. This adjusts the levels of carbon dioxide and oxygen in the blood.

> Carbon dioxide receptors are found in the medulla of the brain (see Chapter 5), and in the aorta and carotid arteries leading from the heart (Chapter 4).

The effects of smoking on the lungs and associated tissues

If the lungs are to be able to exchange gases properly, the air passages need to be clear, the alveoli to be free from dirt particles and bacteria, and they must have as big a surface area as possible in contact with the blood. There is one habit that can upset all of these conditions – smoking.

Links between smoking and diseases of the lungs are now a proven fact. Smoking is associated with lung cancer, bronchitis and emphysema. It is also a major contributing factor to other problems, such as coronary heart disease and ulcers of the stomach and duodenum (part of the intestine). Pregnant women who smoke are more likely to give birth to underweight babies. We need to deal with some of these effects in more detail.

Effects of smoke on the lining of the air passages

You saw above how the lungs are kept free of particles of dirt and bacteria by the action of mucus and cilia. In the trachea and bronchi of a smoker, the cilia are destroyed by the chemicals in cigarette smoke. Compare Figure 2.4 with the same photo taken of the lining of a smoker's airways (Figure 2.11).

The reduced numbers of cilia mean that the mucus is not swept away from the lungs, but remains to clog the air passages. This is made worse by the fact that the smoke irritates the lining of the airways, stimulating the cells to secrete more mucus. The clogging mucus is the source of 'smoker's cough'. Irritation of the bronchial tree, along with infections from bacteria in the mucus can cause the lung disease **bronchitis**. Bronchitis blocks normal air flow, so the sufferer has difficulty breathing properly.

Emphysema

Emphysema is another lung disease that kills about 20 000 people in Britain every year. Smoke damages the walls of the alveoli, which break down and fuse together again, forming enlarged, irregular air spaces (Figure 2.12).

This greatly reduces the surface area for gas exchange, which becomes very inefficient. The blood of a person with emphysema carries less oxygen. In serious cases, this leads to the sufferer being unable to carry out even mild exercise, such as walking. Emphysema patients often have to have a supply of oxygen nearby at all times (Figure 2.13). There is no cure for emphysema, and usually the sufferer dies after a long and distressing illness.

Figure 2.11 *This electron micrograph shows cilia from the trachea of a smoker. Notice the reduced numbers of cilia compared with a normal trachea (see Figure 2.4).*

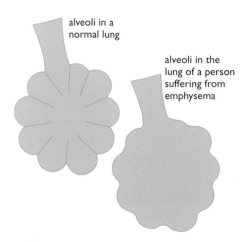

alveoli in a normal lung

alveoli in the lung of a person suffering from emphysema

Figure 2.12 *The alveoli of a person suffering from emphysema have a greatly reduced surface area and inefficient gas exchange.*

Figure 2.13 *This man suffers from emphysema and has to breathe air enriched with oxygen to stay alive.*

Lung cancer

Evidence of the link between smoking and lung cancer first appeared in the 1950s. In one study, a number of patients in hospital were given a series of questions about their lifestyles. They were asked about their work, hobbies, housing and so on, including a question about how many cigarettes they smoked. The same questionnaire was given to two groups of patients. The first group were all suffering from lung cancer. The second, **control** group were in hospital with various other illnesses, but not lung cancer. To make it a fair comparison, the control patients were matched with the lung cancer patients for sex, age and so on.

When the results were compared, one difference stood out (Table 2.2). A greater proportion of the lung cancer patients were smokers than in the control patients. There seemed to be a connection between smoking and getting lung cancer.

Patients	Percentage of patients who were non-smokers	Percentage of patients who smoked more than 15 cigarettes a day
with lung cancer	0.5	25
control (with illnesses other than lung cancer)	4.5	13

Table 2.2: *Comparison of the smoking habits of lung cancer patients and other patients.*

Although the results didn't prove that smoking caused lung cancer, there was a statistically significant link between smoking and the disease: this is called a 'correlation'.

Over 20 similar investigations in nine countries have revealed the same findings. In 1962 a report called 'Smoking and health' was published by the Royal College of Physicians of London, which warned the public about the dangers of smoking. Not surprisingly, the first people to take the findings seriously were doctors, many of whom stopped smoking. This was reflected in their death rates from lung cancer. In 10 years, while deaths among the general male population had risen by 7%, the deaths of male doctors from the disease had *fallen* by 38%.

Cigarette smoke contains a strongly addictive drug – **nicotine**. It also contains at least 17 chemicals that are known to cause cancer. These chemicals are called **carcinogens**, and are contained in the **tar** that collects in a smoker's lungs. Cancer happens when cells mutate and start to divide uncontrollably, forming a **tumour** (Figure 2.14). If a lung cancer patient is lucky, he or she may have the tumour removed by an operation before the cancer cells spread to other tissues of the body. Unfortunately tumours in the lungs usually cause no pain, so they are not discovered until it is too late – it may be inoperable, or tumours may have developed elsewhere.

If you smoke you are not *bound* to get lung cancer, but the risks that you will get it are much greater. In fact, the more cigarettes you smoke, the more the risk that you will get the disease increases (Figure 2.15).

The obvious thing to do is not to start smoking. However, if you are a smoker, giving up the habit soon improves your chance of survival (Figure 2.16). After a few years, the likelihood of your dying from a smoking-related disease is almost back to the level of a non-smoker.

People often talk about 'yellow nicotine stains'. In fact it is the *tar* that stains a smoker's fingers and teeth. Nicotine is a colourless, odourless chemical.

Figure 2.14 *This lung is from a patient with lung cancer.*

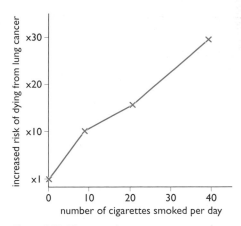

Figure 2.15 *The more cigarettes a person smokes, the more likely it is they will die of lung cancer. For example, smoking 20 cigarettes a day increases the risk by about 15 times.*

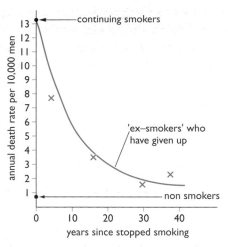

Figure 2.16 *Death rates from lung cancer for smokers, non-smokers and ex-smokers.*

Studies have shown that the type of cigarette smoked makes very little difference to the smoker's risk of getting lung cancer. Filtered and 'low tar' cigarettes only reduce the risk slightly.

Carbon monoxide in smoke

One of the harmful chemicals in cigarette smoke is the poisonous gas **carbon monoxide**. When this gas is breathed in with the smoke, it enters the bloodstream and interferes with the ability of the blood to carry oxygen. Oxygen is carried around in the blood in the red blood cells, attached to a chemical called **haemoglobin** (see Chapter 4). Carbon monoxide can combine with the haemoglobin much more tightly than oxygen can, forming a compound called **carboxyhaemoglobin**. The haemoglobin will combine with carbon monoxide in preference to oxygen. When this happens, the blood carries much less oxygen around the body. Carbon monoxide from smoking is also a major cause of heart disease (Chapter 4).

If a pregnant woman smokes, she will be depriving her unborn fetus of oxygen (Figure 2.17). This has an effect on its growth and development, and leads to the mass of the baby at birth being lower, on average, than the mass of babies born to non-smokers.

Figure 2.17 *Smoking during pregnancy affects the growth and development of the baby.*

Some recent smoking statistics

- 10 million adults in the UK smoke cigarettes – 21% of women and 22% of men. Smoking in women is highest (31%) in the 20–24 age group. It is highest (30%) in men aged 25–34.

- Two-thirds of UK smokers start before age 18. In England, one in seven 15 year olds are regular smokers: 11% of boys and 17% of girls.

- 22% of women and 30% of men in the UK are now ex-smokers. Surveys show that about two-thirds of current smokers would like to stop smoking.

- In the UK, smoking kills around 114 000 people every year. About half of all regular cigarette smokers will eventually be killed by their addiction.

- Smoking causes almost 90% of deaths from lung cancer, around 80% of deaths from bronchitis and emphysema, and around 17% of deaths from heart disease.

Some recent smoking statistics *cont.*

- In 2008–09 the UK Government earned £9700 million in revenue from tax on tobacco. It spent less than £100 million on education campaigns and other help for people to stop smoking.

- The World Health Organization estimates that there are around 1.3 billion smokers in the world, of whom almost 1 billion are men. This represents about one third of the global population aged 15 and over. The vast majority of these people, around 84% or 1 billion people, live in developing countries.*

- Tobacco kills more people than AIDS, legal and illegal drugs, road accidents, murder and suicide combined. Currently around 5 million people worldwide die each year from tobacco-related causes. In comparison, HIV/AIDS is responsible for 3 million deaths per year.*

- In China alone, there are about 350 million smokers. If the current smoking patterns in China continue, around 100 million Chinese men now aged under 29 will die as a result of their tobacco use.*

*Sources: Action on Smoking and Health (ASH) fact sheet (January 2010); *ASH research report (August 2007).*

Giving up smoking

Most smokers admit that they would like to find a way to give up the habit. The trouble is that the nicotine in tobacco is a very addictive drug, and causes withdrawal symptoms when people stop smoking. These include cravings for a cigarette, restlessness and a tendency to put on weight (nicotine depresses the appetite).

There are various ways that smokers can be helped to 'kick the habit'. One of the most successful methods is the use of nicotine patches (Figure 2.18) or nicotine chewing gum. These provide the smoker who is trying to give up with a source of nicotine, without the harmful tar of cigarettes. The nicotine is absorbed through the skin (with patches) or through the mouth (from gum) and reduces the craving for a cigarette. Gradually the 'ex-smoker' reduces the nicotine dose until they are weaned off the habit.

There are several other ways that people use to help them give up smoking, including the use of drugs that reduce withdrawal symptoms, acupuncture and even hypnotism.

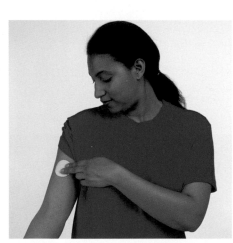

Figure 2.18 *Increasing numbers of young women are smokers. Many turn to nicotine patches to help them give up.*

You could carry out an Internet search to find out about the different methods people use to help them give up smoking. Which methods have the highest success rate?

End of Chapter Checklist

You should now be able to:

✓ recall the structure of the thorax, including the ribs, intercostal muscles, diaphragm, trachea, bronchi, bronchioles, alveoli and pleural membranes

✓ explain the role of the intercostal muscles and the diaphragm in ventilation

✓ explain how alveoli are adapted for gas exchange

✓ describe a simple experiment to investigate the effect of exercise on breathing rate

✓ describe a simple experiment to compare the amount of carbon dioxide in inhaled and exhaled air

✓ explain breathing movements and lung capacities as shown by a spirometer trace, including the terms tidal volume and vital capacity

✓ explain how the levels of carbon dioxide and oxygen in the blood are regulated

✓ understand the biological consequences of smoking on the lungs and the circulatory system.

Questions

1 Copy and complete the table, which shows what happens in the thorax during ventilation of the lungs. Two boxes have been completed for you.

	Action during inhalation	**Action during exhalation**
external intercostal muscles	contract	
internal intercostal muscles		
ribs		move down and in
diaphragm		
volume of thorax		
pressure in thorax		
volume of air in lungs		

2 A student wrote the following about the lungs.

When we breathe in, our lungs inflate, sucking air in and pushing the ribs up and out, and forcing the diaphragm down. This is called respiration. In the air sacs of the lungs the air enters the blood. The blood then takes the air around the body, where it is used by the cells. The blood returns to the lungs to be cleaned. When we breathe out, our lungs deflate, pulling the diaphragm up and the ribs down. The stale air is pushed out of the lungs.

The student did not have a good understanding of the workings of the lungs. Rewrite her description, using correct biological words and ideas.

3 Sometimes, people injured in an accident such as a car crash suffer from a *pneumothorax*. This is an injury where the chest wall is punctured, allowing air to enter the pleural cavity (see Figure 2.1). A patient was brought to the casualty department of a hospital, suffering from a pneumothorax on the left side of his chest. His left lung had collapsed, but he was able to breathe normally with his right lung.

a) Explain why a pneumothorax caused the left lung to collapse.

b) Explain why the right lung was not affected.

c) If a patient's lung is injured or infected, a surgeon can sometimes 'rest' it by performing an operation called an *artificial pneumothorax*. What do you think might be involved in this operation?

4 Briefly explain the importance of the following.

a) The trachea wall contains C-shaped rings of cartilage.

b) The distance between the air in an alveolus and the blood in an alveolar capillary is less than 1/1000th of a millimetre.

c) The lining of the trachea contains mucus-secreting cells and cells with cilia.

d) Smokers have a lower concentration of oxygen in their blood than non-smokers.

e) Nicotine patches and nicotine chewing gum can help someone give up smoking.

f) The lungs have a surface area of about 60 m^2 and a good blood supply.

5 The table shows the percentage of gases in inhaled and exhaled air.

Gas	Inhaled air %	Exhaled air %
nitrogen	78	79
oxygen		
carbon dioxide		
other gases (mainly argon)	1	1

a) Copy the table and fill in the gaps by choosing from the following numbers:

21   4   0.04   16

b) Explain why the percentage of carbon dioxide is so different.

c) Explain why exhaling is a form of excretion.

d) The following features can be seen in the lungs:

i) thin membranes between the alveoli and the blood supply

ii) a good blood supply

iii) a large surface area.

In each case explain how the feature helps gas exchange to happen quickly.

6 Explain the differences between the lung diseases bronchitis and emphysema.

7 A long-term investigation was carried out into the link between smoking and lung cancer. The smoking habits of male doctors aged 35 or over were determined while they were still alive, then the number and causes of deaths among them were monitored over a number of years. (Note that this survey was carried out in the 1950s – very few doctors smoke these days!) The results are shown in the graph.

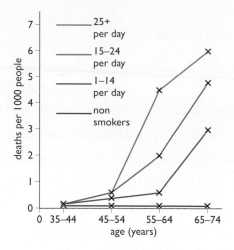

a) Write a paragraph to explain what the researchers found out from the investigation.

b) How many deaths from lung cancer would be expected for men aged 55 who smoked 25 cigarettes a day up until their death? How many deaths from lung cancer would be expected for men in the same age group smoking 10 a day?

c) Table 2.2 (page 28) shows the findings of another study linking lung cancer with smoking. Which do you think is the more convincing evidence of the link, this investigation or the findings illustrated in Table 2.2?

8 Design and make a hard-hitting leaflet explaining the link between smoking and lung cancer. It should be aimed at encouraging an adult smoker to give up the habit. You could use a suitable computer software package to produce your design. Include some smoking statistics, perhaps from an Internet search. However, don't use too many, or they may put the person off reading the leaflet!

Chapter 3: Food and Digestion

We need food for three main reasons:

- to supply us with a 'fuel' for energy
- to provide materials for growth and repair of tissues
- to help fight disease and keep our bodies healthy.

A balanced diet

The food that we eat is called our **diet**. No matter what you like to eat, if your body is to work properly and stay healthy, your diet must include five groups of food substances – **carbohydrates**, **lipids**, **proteins**, **minerals** and **vitamins** – as well as **water** and **fibre**. Food should provide you with all of these substances, but they must also be present in the *right* amounts. A diet that provides enough of these substances and in the correct proportions to keep you healthy is called a **balanced diet** (Figure 3.1). We will deal with each type of food in turn, to find out about its chemistry and the role that it plays in the body.

Food is essential for life. The nutrients obtained from it are used in many different ways by the body. This chapter looks at the different kinds of food, and how the food is broken down by the digestive system and absorbed into the blood, so that it can be carried to all the tissues of the body.

Figure 3.1 *A balanced diet contains all the types of food the body needs, in just the right amounts.*

Carbohydrates

Carbohydrates only make up about 5% of the mass of the human body, but they have a very important role. They are the body's main 'fuel' for supplying cells with energy. Cells release this energy by oxidising a sugar called **glucose**, in the process called cell respiration (see Chapter 1). Glucose and other sugars are one sort of carbohydrate.

Glucose is found naturally in many sweet-tasting foods, such as fruits and vegetables. Other foods contain different sugars, such as the fruit sugar called **fructose**, and the milk sugar, **lactose**. Ordinary table sugar, the sort some people put in their tea or coffee, is called **sucrose**. Sucrose is the main sugar that is

The chemical formula for glucose is $C_6H_{12}O_6$. Like all carbohydrates, glucose contains only the elements carbon, hydrogen and oxygen. The 'carbo' part of the name refers to carbon, and the 'hydrate' part refers to the fact that the hydrogen and oxygen atoms are in the ratio two to one, as in water (H_2O).

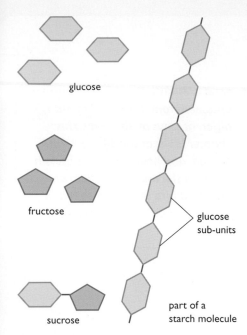

Figure 3.2 *Glucose and fructose are 'single sugar' molecules. A molecule of glucose joined to a molecule of fructose forms the 'double sugar' called sucrose. Starch is a polymer of many glucose sub-units.*

'Single' sugars such as glucose and fructose are called **monosaccharides**. Sucrose molecules are made of two monosaccharides (glucose and fructose) joined together, so sucrose is called a **disaccharide**. Lactose is also a disaccharide of glucose joined to a monosaccharide called galactose. Polymers of sugars, such as starch and glycogen, are called **polysaccharides**.

Fats and oils are both known as **lipids**.

transported through plant stems. This is why we can extract it from sugar cane, which is the stem of a large grass-like plant. Sugars have two physical properties that you will probably know: they all taste sweet, and they are all soluble in water.

We can get all the sugar we need from natural foods such as fruits and vegetables, and from the digestion of starch. Many 'processed' foods contain large amounts of *added* sugar. For example, a typical can of cola can contain up to 27 g, or seven teaspoonfuls! There is hidden sugar in many other foods. A tin of baked beans contains about 10 g of added sugar. This is on top of all the food that we eat with a more obvious sugar content, such as cakes, biscuits and sweets. One of the health problems resulting from all this sugar in our diet is **tooth decay** (see page 43).

In fact, we get most of the carbohydrate in our diet not from sugars, but from **starch**. Starch is a large, *insoluble* molecule. Because it does not dissolve, it is found as a storage carbohydrate in many plants, such as potato, rice, wheat and millet. The 'staple diets' of people from around the world are starchy foods like rice, potatoes, bread and pasta. Starch is made up of long chains of hundreds of glucose molecules joined together. It is called a **polymer** of glucose (Figure 3.2).

Starch is only found in plant tissues, but animal cells sometimes contain a very similar carbohydrate called **glycogen**. This is also a polymer of glucose, and is found in tissues such as liver and muscle, where it acts as a store of energy for these organs.

As you will see, large carbohydrates such as starch and glycogen have to be broken down into simple sugars during digestion, so that they can be absorbed into the blood.

Another carbohydrate that is a polymer of glucose is **cellulose**, the material that makes up plant cell walls. Humans are *not* able to digest cellulose, because our gut doesn't make the enzyme needed to break down the cellulose molecule. This means that we are not able to use cellulose as a source of energy. However, it still has a vitally important function in our diet. It forms **dietary fibre** or **roughage**, which gives the muscles of the gut something to push against as the food is moved through the intestine (see 'peristalsis', page 44). This keeps the gut contents moving, avoiding constipation and helping to prevent serious diseases of the intestine, such as colitis and bowel cancer.

Lipids (fats and oils)

Lipids contain the same three elements as carbohydrates – carbon, hydrogen and oxygen – but the proportion of oxygen in a lipid is much lower than in a carbohydrate. For example, beef and lamb both contain a fat called tristearin, which has the formula $C_{51}H_{98}O_6$. This fat, like other animal fats, is a solid at room temperature, but melts if you warm it up. On the other hand, plant lipids are usually liquid at room temperature, and are called **oils**. Meat, butter, cheese, milk, eggs and oily fish are all rich in animal fats, as well as foods fried in fat or oil, such as chips. Plant oils include many types used for cooking, such as olive oil, corn oil and rapeseed oil, as well as products made from oils, such as margarine (Figure 3.3).

Lipids make up about 10% of our body's mass. They form an essential part of the structure of all cells, and fat is deposited in certain parts of the body as a long-term store of energy, for example under the skin and around the heart and kidneys. The fat layer under the skin acts as insulation, reducing heat loss through the surface of the body. Fat around organs such as the kidneys also helps to protect them from mechanical damage.

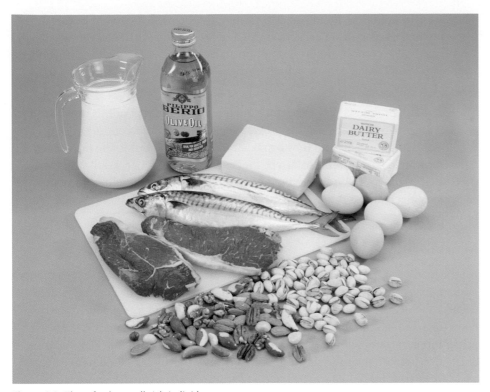

Figure 3.3 *These foods are all rich in lipids.*

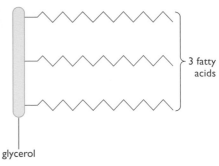

glycerol

Figure 3.4 *Lipids are made up of a molecule of glycerol joined to three fatty acids. The many different fatty acids form the variable part of the molecule.*

The chemical 'building blocks' of lipids are two types of molecule called **glycerol** and **fatty acids**. Glycerol is an oily liquid. It is also known as glycerine, and is used in many types of cosmetics. In lipids, a molecule of glycerol is joined to three fatty acid molecules. There are a large number of different fatty acid molecules, which gives us the many different kinds of lipid found in food (Figure 3.4).

Although lipids are an essential part of our diet, too much lipid is unhealthy, especially a type called **saturated** fat, and a lipid compound called **cholesterol**. These substances have been linked to heart disease (see Chapter 4).

Saturated lipids (fats) are more common in food from animal sources, such as meat and dairy products. 'Saturated' is a word used in chemistry, which means that the fatty acids of the lipids contain no double bonds. Other lipids are **unsaturated**, which means that their fatty acids contain double bonds. These are more common in plant oils. There is evidence that unsaturated lipids are healthier for us than saturated ones.

Cholesterol is a substance that the body gets from food such as eggs and meat, but we also make cholesterol in our liver. It is an essential part of all cells, but too much cholesterol causes heart disease.

Proteins

Proteins make up about 18% of the mass of the body. This is the second largest fraction after water. All cells contain protein, so we need it for growth and repair of tissues. Many compounds in the body are made from protein, including enzymes.

Most foods contain some protein, but certain foods such as meat, fish, cheese and eggs are particularly rich in it. You will notice that these foods are animal products. Plant material generally contains less protein, but some foods, especially beans, peas and nuts, are richer in protein than others.

However, we don't need much protein in our diet to stay healthy. Doctors recommend a maximum daily intake of about 70 g. In more economically developed countries, people often eat far more protein than they need, whereas in many poorer countries a protein-deficiency disease called **kwashiorkor** is common (Figure 3.5).

Like starch, proteins are also polymers, but whereas starch is made from a single molecular building block (glucose), proteins are made from 20 different sub-units

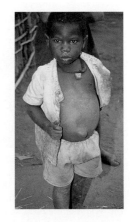

Figure 3.5 *This child is suffering from a lack of protein in his diet, a disease called kwashiorkor. His swollen belly is not due to a full stomach, but is caused by fluid collecting in the tissues. Other symptoms include loss of weight, poor muscle growth, general weakness and flaky skin.*

Humans can make about half of the 20 amino acids that they need, but the other 10 have to be taken in as part of the diet. These 10 are called **essential amino acids**. There are higher amounts of essential amino acids in meat, fish, eggs and dairy products. If you are a vegetarian, you can still get all the essential amino acids you need, as long as you eat a varied diet that includes a range of different plant materials.

(a)

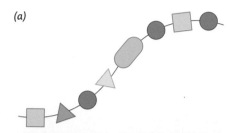

(b)

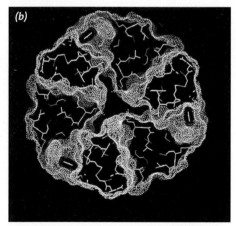

Figure 3.6 *(a) A chain of amino acids forming part of a protein molecule. Each shape represents a different amino acid. (b) A computer model of the protein insulin. This substance, like all proteins, is made of a long chain of amino acids arranged in a specific order.*

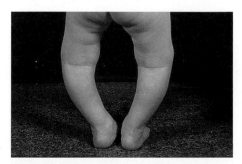

Figure 3.7 *The legs of this child show the symptoms of rickets. This is due to lack of calcium or a lack of vitamin D in the diet, leading to poor bone growth. The bones stay soft and can't support the weight of the body, so they become deformed.*

called **amino acids**. All amino acids contain four chemical elements: carbon, hydrogen and oxygen (as in carbohydrates and fats) along with nitrogen. Two amino acids also contain sulfur. The amino acids are linked together in long chains, which are usually folded up or twisted into spirals, with cross-links holding the chains together (Figure 3.6).

The *shape* of a protein is very important in allowing it to carry out its function, and the *order* of amino acids in the protein decides its shape. Because there are 20 different amino acids, and they can be arranged in any order, the number of different protein structures that can be made is enormous. As a result, there are thousands of different kinds of proteins in organisms, from structural proteins such as collagen and keratin in skin and nails, to proteins with more specific functions, such as enzymes and haemoglobin.

Minerals

All the foods you have read about so far are made from just five chemical elements: carbon, hydrogen, oxygen, nitrogen and sulfur. Our bodies contain many other elements which we get from our food. Some are present in large amounts in the body, for example calcium, which is used for making teeth and bones. Others are present in much smaller amounts, but still have essential jobs to do. For instance our bodies contain about 3 g of iron, but without it our blood would not be able to carry oxygen. Table 3.1 shows just a few of these elements and the reasons they are needed. They are called **minerals** or **mineral elements**.

Mineral	Approximate mass in an adult body (g)	Location or role in body	Examples of foods rich in minerals
calcium	1000	making teeth and bones	dairy products, fish, bread, vegetables
phosphorus	650	making teeth and bones; part of many chemicals, e.g. DNA	most foods
sodium	100	in body fluids, e.g. blood	common salt, most foods
chlorine	100	in body fluids, e.g. blood	common salt, most foods
magnesium	30	making bones; found inside cells	green vegetables
iron	3	part of haemoglobin in red blood cells, helps carry oxygen	red meat, liver, eggs, some vegetables, e.g. spinach

Table 3.1: *Some examples of minerals needed by the body.*

If a person doesn't get enough of a mineral from their diet, they will show the symptoms of a **mineral deficiency disease**. For example, a 1-year-old child needs to consume about 0.6 g (600 mg) of calcium every day, to make the bones grow properly and harden. Anything less than this over a prolonged period could result in poor bone development. The bones will become deformed, resulting in a disease called **rickets** (Figure 3.7). Rickets can also be caused by lack of vitamin D in the diet (see below).

Similarly, 16-year-olds need about 12 mg of iron in their daily food intake. If they don't get this amount, they can't make enough haemoglobin for their red blood cells (see Chapter 4). This causes a condition called **anaemia**. People who are anaemic become tired and lack energy, because their blood doesn't carry enough oxygen.

Vitamins

During the early part of the twentieth century, experiments were carried out that identified another class of food substances. When young laboratory rats were fed a diet of pure carbohydrate, lipid and protein, they all became ill and died. If they were fed on the same pure foods with a little added milk, they grew normally. The milk contained chemicals that the rats needed in small amounts to stay healthy. These chemicals are called **vitamins**. The results of one of these experiments are shown in Figure 3.8.

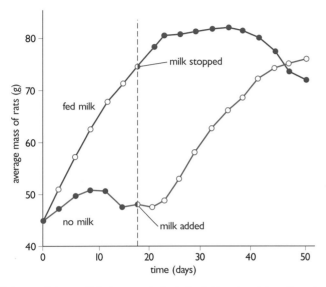

Figure 3.8 *Rats were fed on a diet of pure carbohydrate, lipid and protein, with and without added milk. Vitamins in the milk had a dramatic effect on their growth.*

At first, the chemical nature of vitamins was not known, and they were given letters to distinguish between them, such as vitamin A, vitamin B and so on. Each was identified by the effect a lack of the vitamin, or **vitamin deficiency**, had on the body. For example, **vitamin D** is needed for growing bones to take up calcium salts. A deficiency of this vitamin can result in rickets (Figure 3.7), just as a lack of calcium can.

We now know the chemical structure of the vitamins and the exact ways in which they work in the body. As with vitamin D, each has a particular function. **Vitamin A** is needed to make a light-sensitive chemical in the retina of the eye (see Chapter 5). A lack of this vitamin causes **night blindness**, where the person finds it difficult to see in dim light. **Vitamin C** is needed to make fibres of a material called connective tissue. This acts as a 'glue', bonding cells together in a tissue. It is found in the walls of blood vessels and in the skin and lining surfaces of the body. Vitamin C deficiency leads to a disease called **scurvy**, where wounds fail to heal, and bleeding occurs in various places in the body. This is especially noticeable in the gums (Figure 3.9).

Vitamin B is not a single substance, but a collection of many different substances called the vitamin B group. It includes **vitamins B1 (thiamine), B2 (riboflavin)** and **B3 (niacin)**. These compounds have roles in helping with the process of cell respiration. A different deficiency disease is produced if any of them is lacking from the diet. For example, lack of vitamin B1 results in weakening of the muscles and paralysis, a disease called **beri-beri**.

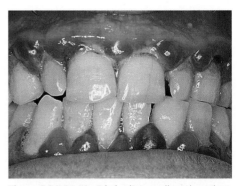

Figure 3.9 *Vitamin C helps lining cells such as those in the mouth and gums stick to each other. Lack of vitamin C causes scurvy, where the mouth and gums become damaged and bleed.*

The cure for scurvy was discovered as long ago as 1753. Sailors on long voyages often got scurvy because they ate very little fresh fruit and vegetables (the main source of vitamin C). A ship's doctor called James Lind wrote an account of how the disease could quickly be cured by eating fresh oranges and lemons. The famous explorer Captain Cook, on his world voyages in 1772 and 1775, kept his sailors healthy by making sure that they ate fresh fruit. By 1804, all British sailors were made to drink lime juice to prevent scurvy. This is how they came to be called 'limeys', a word that was later used by Americans for all British people.

The main vitamins, their role in the body and some foods which are good sources of each, are summarised in Table 3.2.

Notice that the amounts of vitamins that we need are very small, but we cannot stay healthy without them.

Vitamin	Recommended daily amount in diet[1]	Use in the body	Effect of deficiency	Some foods that are a good source of the vitamin
A	0.8 mg	making a chemical in the retina; also protects the surface of the eye	night blindness, damaged cornea of eye	fish liver oils, liver, butter, margarine, carrots
B1	1.4 mg	helps with cell respiration	beri-beri	yeast extract, cereals
B2	1.6 mg	helps with cell respiration	poor growth, dry skin	green vegetables, eggs, fish
B3	18 mg	helps with cell respiration	pellagra (dry red skin, poor growth, and digestive disorders)	liver, meat, fish.
C	60 mg	sticks together cells lining surfaces such as the mouth	scurvy	fresh fruit and vegetables
D	5 µg	helps bones absorb calcium and phosphate	rickets, poor teeth	fish liver oils; also made in skin in sunlight

[1]Figures are the European Union's recommended daily intake for an adult (1993). 'mg' stands for milligram (a thousandth of a gram) and 'µg' for microgram (a millionth of a gram).

Table 3.2: *Summary of the main vitamins.*

Food tests

It is possible to carry out simple tests to find out if a food contains starch, glucose, protein or lipid. The following descriptions use pure food samples for the tests, but it is possible to do them on normal foods too. Unless the food is a liquid, like milk, it needs to be cut up into small pieces and ground with a pestle and mortar, then shaken with some water in a test tube. This is done to extract the components of the food and dissolve any soluble substances, such as sugars.

Experiment 6

Test for starch

A little starch is placed on a spotting tile. A drop of yellow-brown **iodine solution** is added to the starch. The iodine reacts with the starch, forming a very dark blue, or 'blue-black' colour (Figure 3.10a). Starch is insoluble, but this test will work on a solid sample of food, such as potato, or a suspension of starch in water.

Test for glucose

Glucose is called a **reducing sugar**. This is because the test for glucose involves reducing an alkaline solution of copper(II) sulfate to copper(I) oxide.

A small spatula measure of glucose is placed in a test tube and a little water added (about 2 cm deep). The tube is shaken to dissolve the glucose. Several drops of **Benedict's solution** are added to the tube, enough to colour the mixture blue (Figure 3.10b1).

A water bath is prepared by half-filling a beaker with water and heating it on a tripod and gauze. The test tube is placed in the beaker and the water allowed to boil (using a water bath is safer than heating the tube directly in the Bunsen burner). After a few seconds the clear blue solution gradually changes colour, forming a cloudy orange or 'brick red' precipitate of copper(I) oxide (Figure 3.10b2).

All other 'single' sugars, such as fructose, are reducing sugars, as well as some 'double' sugars, such as the milk sugar, lactose. However, ordinary table sugar (sucrose) is not. If sucrose is boiled with Benedict's solution it will stay a clear blue colour.

Test for protein

The test for protein is sometimes called the 'biuret' test, after the coloured compound that is formed.

A little protein, such as powdered egg white (albumen), is placed in a test tube and about 2 cm depth of water added. The tube is shaken to mix the powder with the water (Figure 3.10c1). An equal volume of dilute (5%) **potassium hydroxide** solution is added and the tube shaken again. Finally two drops of 1% **copper sulfate** solution are added. A mauve colour develops (Figure 3.10c2). (Sometimes these two solutions are supplied already mixed together as 'biuret solution'.)

Test for lipid

Fats and oils are insoluble in water, but will dissolve in ethanol (alcohol). The test for lipid uses this fact.

A pipette is used to place one drop of olive oil in the bottom of a test tube. About 2 cm depth of ethanol is added, and the tube is shaken to dissolve the oil. The solution is poured into a test tube that is about three-quarters full with cold water. A white cloudy layer forms on the top of the water (Figure 3.10d). The white layer is caused by the ethanol dissolving in the water and leaving the lipid behind as a suspension of tiny droplets, called an emulsion.

Energy from food

Some foods contain more energy than others. It depends on the proportions of carbohydrate, lipid and protein that they contain. Their energy content is measured in **kilojoules** (**kJ**). If a gram of carbohydrate is fully oxidised, it produces about 17 kJ, whereas a gram of lipid yields over twice as much as this (39 kJ). Protein can produce about 18 kJ. If you look on a food label, it usually shows the energy content of the food, along with the amounts of different nutrients that it contains (Figure 3.11).

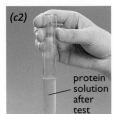
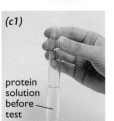
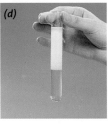

Figure 3.10 Results of tests for (a) starch, (b) glucose, (c) protein and (d) fat.

Figure 3.11 Food packaging is labelled with the proportions of different food types that it contains, along with its energy content. The energy in units called kilocalories (kcal) is also shown, but scientists no longer use this old-fashioned unit.

Foods with a high percentage of lipid, such as butter or nuts, contain a large amount of energy. Others, like fruits and vegetables, which are mainly composed of water, have a much lower energy content (Table 3.3).

Food scientists measure the amount of energy in a sample of food by burning it in a calorimeter (Figure 3.12). The calorimeter is filled with oxygen, to make sure that the food will burn easily. A heating filament carrying an electrical current ignites the food. The energy given out by the burning food is measured by using it to heat up water flowing through a coil in the calorimeter.

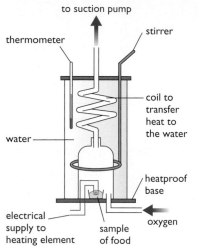

Figure 3.12 *A food calorimeter.*

If you have samples of food that will easily burn in air, you can measure the energy in them by a similar method, using the heat from the burning food to warm up water in a test tube (see Experiment 7 on the next page).

Food	kJ per 100 g
margarine	3 200
butter	3 120
peanuts	2 400
samosa	2 400
chocolate	2 300
Cheddar cheese	1 700
grilled bacon	1 670
table sugar	1 650
grilled pork sausages	1 550
cornflakes	1 530
rice	1 500
spaghetti	1 450

Food	kJ per 100 g
fried beefburger	1 100
white bread	1 060
chips	990
grilled beef steak	930
fried cod	850
roast chicken	770
boiled potatoes	340
milk	270
baked beans	270
yoghurt	200
boiled cabbage	60
lettuce	40

Table 3.3: *Energy content of some common foods.*

Even while you are asleep you need a supply of energy, for keeping warm, for your heartbeat, to allow messages to be sent through your nerves, and for other body functions. However, the energy you need at other times depends on the physical work that you do. The total amount of energy that a person needs to keep healthy depends on their age and body size, and also on the amount of activity they do. Table 3.4 shows some examples of how much energy is needed each day by people of different age, sex and occupation.

Age/sex/occupation of person	Energy needed per day (kJ)
newborn baby	2 000
child aged 2	5 000
child aged 6	7 500
girl aged 12–14	9 000
boy aged 12–14	11 000
girl aged 15–17	9 000
boy aged 15–17	12 000
female office worker	9 500
male office worker	10 500
heavy manual worker	15 000
pregnant woman	10 000
breastfeeding woman	11 300

Table 3.4: *The daily energy needs of different types of people.*

Remember that these are approximate figures, and they are averages. Generally, the greater a person's weight, the more energy that person needs. This is why men, with a greater average body mass, need more energy than women. The energy needs of a pregnant woman are increased, mainly because of the extra weight that she has to carry. A heavy manual worker, such as a labourer, needs extra energy for increased muscle activity. Climate also matters. In very cold climates the body will lose more heat, so that a person will need more energy to maintain a constant body temperature, and vice versa in a hot climate. A person's rate of metabolism can be increased under the influence of the hormone thyroxine (see Chapter 6, page 89).

It is not only the recommended energy requirements that vary with age, sex and pregnancy, but also the *content* of the diet. For instance, during pregnancy a woman may need extra iron or calcium in her diet, for the growth of the fetus. In younger women, the blood loss during menstruation (periods) can result in anaemia, producing a need for extra iron in the diet.

Measuring the energy content of a food

If a sample of food will burn well in air, its energy content can be measured using a simplified version of the food calorimeter (Figure 3.13). Suitable foods are dry pasta, crispbread, corn curls or biscuits. It is not advisable to use nuts, since some people are allergic to them.

First of all the mass of the food sample is found, by weighing it on a balance. 20 cm³ of water is placed in a boiling tube, and the tube supported in a clamp on a stand as shown in Figure 3.13. The temperature of the water is measured.

The food is speared on the end of a mounted needle, and then held in a Bunsen burner flame until it catches fire (this may take 30 seconds or so). When the food is alight, the mounted needle is used to hold the burning food underneath the boiling tube of water so that the flame heats up the water. This is continued, relighting the food if it goes out, until the food will no longer burn.

The final temperature of the water is measured, using the thermometer to stir the water gently, to make sure that the heat is evenly distributed.

Two facts are needed to calculate the energy content of the food:

- 4.2 joules of energy raises the temperature of one gram of water by 1 °C.
- 1 cm³ of water has a mass of 1 g.

So multiplying the rise in temperature of the water by the mass of the water and then by 2 gives the number of joules of energy that were transferred to the water. Dividing this by the mass of the food gives the energy/gram:

Energy in joules per gram (J/g) =

$$\frac{(\text{final temperature} - \text{temperature at start}) \times 20\ (\text{g}) \times 4.2\ (\text{J per °C})}{\text{mass of food (g)}}$$

For example, imagine you had a piece of pasta weighing 0.55 g. The starting temperature of the 20 g of water was 21 °C. After using the burning pasta to heat up the water, the temperature of the water was 43 °C.

Energy content of the pasta =

$$\frac{(43 - 21) \times 20 \times 4.2}{0.55} = 3360\ \text{J/g}\ (3400\ \text{J/g to 2 significant figures})$$

Comparison of the energy content of different foods

You may be able to use a similar method to find the energy content of suitable foods that will burn easily. Suggest a hypothesis that you could test about the energy content of the foods, and design an experiment to test your hypothesis. Explain how you will ensure that your results are reliable.

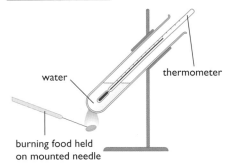

Figure 3.13 *Measuring the energy content of a food sample.*

Taking food into the body through the mouth is known as **ingestion**.

This is a good definition of digestion, useful for exam answers:
'Digestion is the chemical and mechanical breakdown of food. It converts large insoluble molecules into small soluble molecules, which can be absorbed into the blood.'

Digestion

Food, such as a piece of bread, contains carbohydrates, lipids and proteins, but they are not the same carbohydrates, lipids and proteins as in our tissues. The components of the bread must first be broken down into their 'building blocks' before they can be absorbed through the wall of the gut. This process is called **digestion**. The digested molecules – sugars, fatty acids, glycerol and amino acids – along with minerals, vitamins and water, can then be carried around the body in the blood. When they reach the tissues they are reassembled into the molecules that make up our cells.

Digestion is speeded up by **enzymes**, which are biological catalysts (see Chapter 1). Although most enzymes stay inside cells, the digestive enzymes are made by the tissues and glands in the gut, and pass out of cells onto the gut contents, where they act on the food. This **chemical** digestion is helped by **mechanical** digestion. This is the physical breakdown of food. The most obvious place where this happens is in the mouth, where the teeth bite and chew the food, cutting it into smaller pieces that have a larger surface area.

At the same time the food is mixed with saliva, which lubricates it, allowing it to be swallowed more easily. Saliva also contains the enzyme amylase, which starts the breakdown of starch.

Teeth

In mammals, only the lower jaw is movable, while the upper jaw is fused to the skull. Chewing is brought about by the cheek muscles, which move the lower jaw up and down, and allow some side-to-side movement.

As with most mammals, humans have four types of teeth, called incisors, canines, premolars and molars (Figure 3.14). The arrangement of an animal's teeth is called its **dentition.**

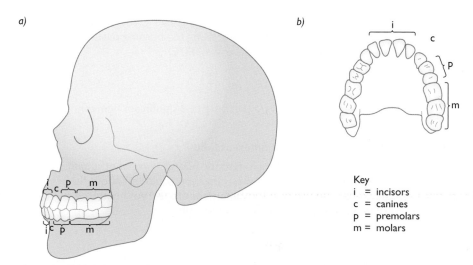

Key
i = incisors
c = canines
p = premolars
m = molars

Figure 3.14 *(a) Side view of human skull and teeth; (b) upper teeth shown from below.*

At the front of the mouth are the **incisors**. They are relatively sharp, chisel-shaped teeth, used for biting off pieces of food. Behind these in each jaw is a pair of **canines** (one on each side of the mouth). In humans the canines are similar in shape to the incisors, and have the same biting function. Behind the canines are the **cheek teeth** (**molars** and **premolars**).

The cheek teeth have a flatter top surface or **crown**, and the top and bottom sets of cheek teeth meet crown to crown, so that they can be used for chewing or crushing food. Human teeth are adapted for dealing with a wide range of food types, from meat and fish to plant roots, stems and leaves.

The crown of a tooth is covered with a non-living material called **enamel** (Figure 3.15), which is the hardest substance in the body. Underneath the enamel is a softer material (but still about as hard as bone) called **dentine**. The middle of the tooth is called the **pulp cavity**. It contains blood vessels and nerves. There are fine channels running through the dentine, filled with cytoplasm. These cytoplasmic strands are kept alive by nutrients and oxygen from the blood vessels in the pulp cavity. The root of the tooth is covered with **cement**, containing **fibres**. This material anchors the tooth in the jawbone but allows a slight degree of movement when the person is chewing.

During their lifetime, mammals have two sets of teeth. The first set is called the **milk** or **deciduous teeth**. In humans these start to grow through the gum when a child is a few months old. By the age of about 21 or 22 months, a child will have 20 teeth, mainly milk teeth. From around the age of 7 years, the milk teeth are pushed out by permanent teeth growing underneath them. The molars at the back are present only as permanent teeth. Eventually a full set of 32 adult teeth is formed (Figure 3.14).

Tooth decay

Tooth decay (dental **caries**) is one of the most common diseases in the world. It is caused by bacteria in the mouth feeding on sugar. The bacteria break down the sugar, forming acids which dissolve the tooth enamel. Once the enamel is penetrated, the acid breaks down the softer dentine underneath. Eventually a cavity is formed in the tooth (Figure 3.16).

Bacteria can then enter this cavity and enlarge it until the decay reaches the nerves in the pulp cavity. Then you feel the pain! These bacteria are also the cause of **periodontal disease**, where the gums become inflamed and so sensitive that they bleed when the teeth are brushed. Periodontal disease can also lead to loss of teeth.

Dental hygiene

Bacteria form an invisible layer on the surface of teeth, called **plaque**. One of the obvious ways you can prevent tooth decay is by regularly brushing your teeth. Plaque takes about 24 hours to form again, so teeth (and gums) should be brushed twice a day. Dentists advise the use of a toothbrush with a small head that will allow the bristles to reach into the crevices between the teeth, or better still an electric toothbrush. Modern electric toothbrushes have a timed cycle, so that you know you have brushed your teeth for the right length of time. You should also use dental floss to clean between the teeth, where it is difficult to brush. If plaque is left on the teeth it soon forms a hard deposit called **tartar**. This has to be removed by a dentist. You should have a check-up at the dentist's every 6 months, even when you have no obvious problems – the dentist may spot a developing tooth or gum problem before it gets worse.

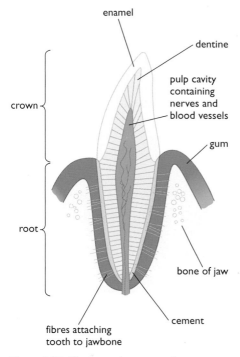

Figure 3.15 *The internal structure of an incisor tooth.*

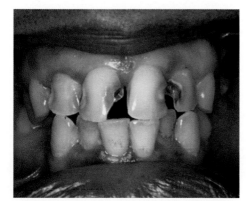

Figure 3.16 *A bad case of tooth decay. One of the causes was too much sugar in the person's diet.*

Fluoride has been shown to reduce tooth decay by strengthening the enamel, especially when teeth are growing. In many places, fluoride is added to drinking water, and has made a big difference to the incidence of decay. If fluoride is not added to your drinking water, you can take fluoride tablets, or use fluoride toothpaste.

Finally, a good balanced diet is essential for teeth to grow healthily. Avoid sweets and sugary drinks, which will reduce the supply of nutrients for the bacteria.

The role of muscles in the gut wall

As well as the teeth, other parts of the gut help with mechanical digestion. For example, the muscles in the wall of the stomach contract to churn up the food while it is being chemically digested.

Muscles are also responsible for moving the food along the gut. The walls of the intestine contain two layers of muscles. One layer has fibres running in rings around the gut. This is the **circular** muscle layer. The other has fibres running down the length of the gut, and is called the **longitudinal** muscle layer. Together these two layers act to push the food along. When the circular muscles contract and the longitudinal muscles relax, the gut is made narrower. When the opposite happens, i.e. the longitudinal muscles contract and the circular muscles relax, the gut becomes wider. Waves of muscle contraction like this pass along the gut, pushing the food along, rather like squeezing toothpaste from a tube (Figure 3.17). This is called **peristalsis**. It means that movement of food in the gut doesn't depend on gravity – we can still eat standing on our heads!

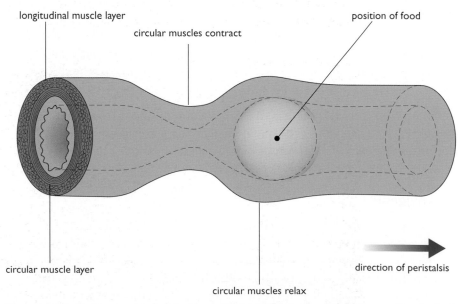

longitudinal muscle layer

circular muscles contract

position of food

circular muscle layer

circular muscles relax

direction of peristalsis

Figure 3.17 *Peristalsis: contraction of circular muscles behind the food narrows the gut, pushing the food along. When the circular muscles are contracted, the longitudinal ones are relaxed, and vice versa.*

Figure 3.18 shows a simplified diagram of the human digestive system. It is simplified so that you can see the order of the organs along the gut. The real gut is much longer than this, and coiled up so that it fills the whole space of the abdomen. Overall, its length in an adult is about 8 m. This gives plenty of time for the food to be broken down and absorbed as it passes through.

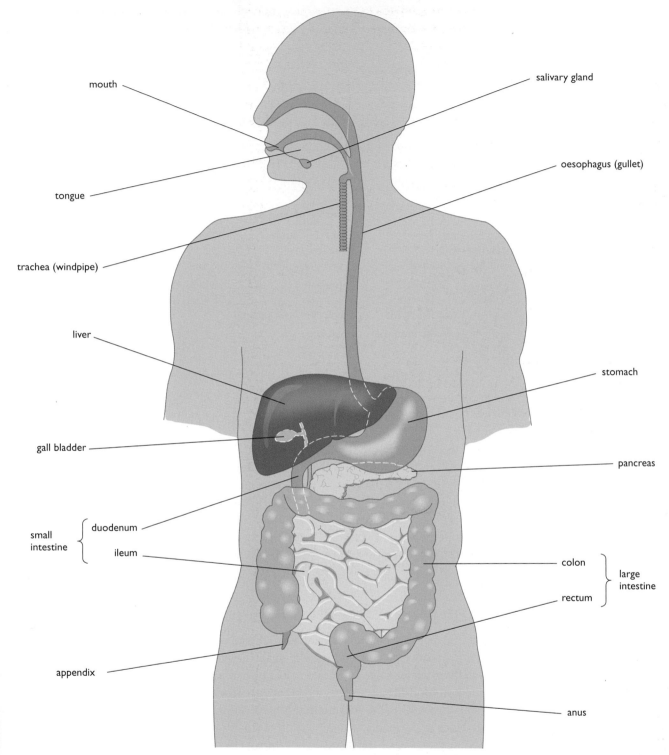

mouth

salivary gland

tongue

oesophagus (gullet)

trachea (windpipe)

liver

stomach

gall bladder

pancreas

small intestine

duodenum

ileum

colon

large intestine

rectum

appendix

anus

Figure 3.18 *The human digestive system.*

The mouth, stomach and the first part of the small intestine (called the **duodenum**) all break down the food using enzymes, either made in the gut wall itself, or by glands such as the **pancreas**. Digestion continues in the last part of the small intestine (the **ileum**) and it is here that the digested food is absorbed. The last part of the gut, the large intestine, is mainly concerned with absorbing water out of the remains, and storing the waste products (**faeces**) before they are removed from the body.

The three main classes of food are broken down by three classes of enzymes. Carbohydrates are digested by enzymes called **carbohydrases**. Proteins are acted upon by **proteases**, and enzymes called **lipases** break down lipids. Some of the places in the gut where these enzymes are made are shown in Table 3.5.

Digestion begins in the mouth. **Saliva** helps moisten the food and contains the enzyme **amylase**, which starts the breakdown of starch. The chewed lump of food, mixed with saliva, then passes along the **oesophagus** (gullet) to the stomach.

Class of enzyme	Examples	Digestive action	Source of enzyme	Where it acts in the gut
carbohydrases	amylase amylase maltase	starch → maltose[1] starch → maltose maltose → **glucose**	salivary glands pancreas wall of small intestine	mouth small intestine small intestine
proteases	pepsin trypsin peptidases	proteins → peptides[2] proteins → peptides peptides → **amino acids**	stomach wall pancreas wall of small intestine	stomach small intestine small intestine
lipases	lipase	lipids → **glycerol** and **fatty acids**	pancreas	small intestine

[1]Maltose is a disaccharide made of two glucose molecules joined together.
[2]Peptides are short chains of amino acids.

Table 3.5: *Some of the enzymes that digest food in the human gut. The substances shown in bold are the end products of digestion that can be absorbed from the gut into the blood.*

Amylase digests starch into maltose. In this reaction, we say that starch is the **substrate** and maltose is the **product**.

The food is held in the stomach for several hours, while initial digestion of protein takes place. The stomach wall secretes **hydrochloric acid**, so the stomach contents are strongly acidic. This has a very important function. It kills bacteria that are taken into the gut along with the food, helping to protect us from food poisoning. The protease enzyme that is made in the stomach, called **pepsin**, has to be able to work in these acidic conditions, and has an optimum pH value of about 2. This is unusually low – most enzymes work best at near neutral conditions (see Chapter 1).

The semi-digested food is held back in the stomach by a ring of muscle at the outlet of the stomach, called a **sphincter** muscle. When this relaxes, it releases the food into the first part of the small intestine, called the **duodenum** (Figure 3.19).

Several digestive enzymes are added to the food in the duodenum. These are made by the **pancreas**, and digest starch, proteins and lipids (Table 3.5). As well as this, the **liver** makes a digestive juice called **bile**. Bile is a green liquid that is stored in the **gall bladder** and passes down the **bile duct** on to the food. Bile does not contain enzymes, but has another important function. It turns any large lipid globules in the food into an emulsion of tiny droplets (Figure 3.20). This increases the surface area of the lipid, so that **lipase** enzymes can break it down more easily.

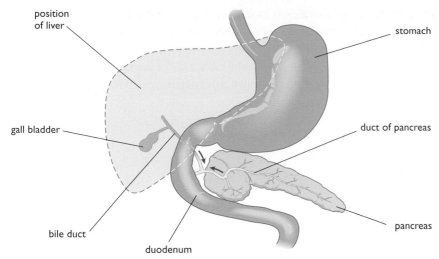

Figure 3.19 *The first part of the small intestine, the duodenum, receives digestive juices from the liver and pancreas through tubes called ducts.*

Bile and pancreatic juice have another function. They are both alkaline. The mixture of semi-digested food and enzymes coming from the stomach is acidic, and needs to be neutralised by the addition of alkali before it continues on its way through the gut.

As the food continues along the intestine, more enzymes are added, until the parts of the food that can be digested have been fully broken down into soluble end products, which can be absorbed. This is the role of the last part of the small intestine, the **ileum**.

Absorption in the ileum

The ileum is highly adapted to absorb the digested food. The lining of the ileum has a very large surface area, which means that it can quickly and efficiently absorb the soluble products of digestion into the blood. The length of the intestine helps to provide a large surface area, and this is aided by folds in its lining, but the greatest increase in area is due to tiny projections from the lining, called **villi** (Figure 3.21).

Figure 3.20 *Bile turns fats into an emulsion of tiny droplets for easier digestion.*

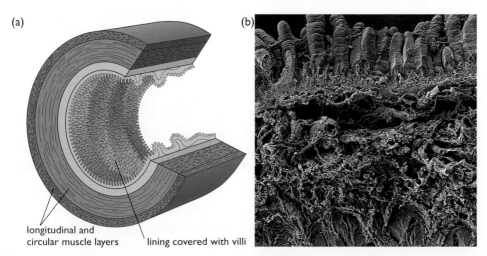

Figure 3.21 *(a) The inside lining of the ileum is adapted to absorb digested food by the presence of millions of tiny villi. (b) A section through the lining, showing the villi.*

The singular of villi is 'villus'. Each villus is only about 1–2 mm long, but there are millions of them, so that the total area of the lining is thought to be about 300 m². This provides a massive area in contact with the digested food. As well as this, high-powered microscopy has revealed that the surface cells of each villus themselves have hundreds of minute projections, called **microvilli**, which increase the surface area for absorption even more (Figure 3.22).

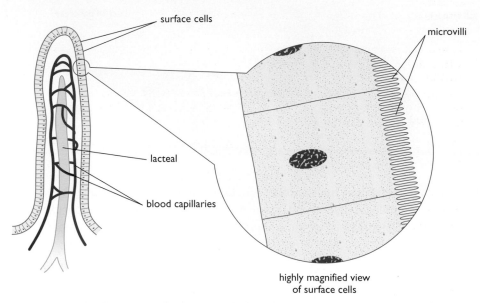

highly magnified view
of surface cells

Figure 3.22 *Each villus contains blood vessels and a lacteal, which absorb the products of digestion. The surface cells of the villus are covered with microvilli, which further increase the surface area for absorption.*

Each villus contains a network of blood capillaries. Most of the digested food enters these blood vessels, but the products of fat digestion, as well as tiny fat droplets, enter a tube in the middle of the villus, called a **lacteal**. The lacteals form part of the body's **lymphatic** system, which transports a liquid called lymph. This **lymph** eventually drains into the blood system too.

The blood vessels from the ileum join up to form a large blood vessel called the **hepatic portal vein**, which leads to the liver (see Chapter 4). The liver acts rather like a food processing works, breaking some molecules down, and building up and storing others. For example, glucose from carbohydrate digestion is converted into **glycogen** and stored in the liver. Later, the glycogen can be converted back into glucose when the body needs it (see Chapter 6).

The digested food molecules are distributed around the body by the blood system (see Chapter 4). The soluble food molecules are absorbed from the blood into cells of tissues, and are used to build new parts of cells. This is called **assimilation**.

The large intestine – elimination of waste

By the time that the contents of the gut have reached the end of the small intestine, most of the digested food, as well as most of the water, has been absorbed. The waste material consists mainly of cellulose (fibre) and other indigestible remains, water, dead and living bacteria and cells lost from the lining of the gut. The function of the first part of the large intestine, called the **colon**, is to absorb most of the remaining water from the contents, leaving a semi-solid waste material called **faeces**. This is stored in the **rectum**, until expelled out of the body through the **anus**.

It has been estimated that the liver has more than 200 different functions to do with metabolism. As well as producing bile, storing glycogen and breaking down hormones, liver cells protect the body by breaking down many poisonous substances (toxins). This is called **detoxification**. One example of detoxification is the removal and breakdown of ethanol (alcohol) from the blood (see page 84).

Removal of faeces by the body is sometimes incorrectly called excretion. Excretion is a word that should only apply to materials that are the waste products of cells of the body. Faeces are not – they consist of waste which has passed through the gut without entering the cells. The correct name for this process is **egestion**.

The effect of temperature on the enzyme trypsin

Trypsin is a protease – an enzyme that digests proteins into short chains of amino acids. It is made by the pancreas (see page 46).

Powdered milk contains a white protein. If milk powder is mixed with water, it forms an opaque suspension. This can be used as a substrate for the trypsin. If a solution of trypsin is added to the milk suspension in a test tube, it will gradually turn clearer. It never turns completely colourless, but eventually it changes from opaque to see-through (translucent).

To find out the effect of temperature on the rate of reaction of trypsin, $5\,cm^3$ of a 1% solution of trypsin and $5\,cm^3$ of a 5% suspension of milk powder are heated separately to a certain temperature in a water bath, and then mixed. The time that the enzyme takes to turn the milk translucent is found. The procedure is then repeated at other temperatures. The rate of reaction is calculated from the time the mixture takes to go translucent.

For example, if the trypsin digested $5\,cm^3$ of milk protein in 2.75 minutes, the rate of reaction in cm^3 per minute is:

$$\text{volume of milk} \div \text{time} = 5 \div 2.75 = 1.82\,cm^3/min$$

A student carried out this experiment at a number of different temperatures. She repeated it three times and calculated the mean time at each temperature and the rate of reaction. Her results are shown in Table 3.6.

Temperature (°C)	Time (min)			Mean time (min)	Mean rate (cm³/min)
	(1)	(2)	(3)		
20	8.28	7.90	8.52	8.23	0.61
30	2.47	3.63	3.23	3.11	1.61
35	2.27	2.20	2.00	2.16	2.31
40	1.67	1.83	1.50	1.67	2.99
45	1.43	1.50	1.43	1.45	3.45
50	1.32	1.33	1.43	1.36	3.68
55	1.42	1.53	1.42	1.46	3.42
60	1.70	1.83	1.90	1.81	2.76
65	2.83	3.33	3.37	3.18	1.57
70	7.58	10.20	9.60	9.13	0.55
80	>30	>30	>30	>30	<0.17

Table 3.6: *Time taken for the enzyme trypsin to digest milk protein at different temperatures. (Columns show three replicates at each temperature, the mean time and mean rate of reaction.)*

Plot a graph of the mean reaction rate against temperature. Can you explain why the curve of the graph is this shape? (See Chapter 1, page 4). You may be able to try a similar experiment yourself.

End of Chapter Checklist

You should now be able to:

✓ recall the chemical elements present in carbohydrates, lipids and proteins

✓ understand the structure of carbohydrates, lipids and proteins as large molecules made up from smaller units: starch and glycogen from simple sugars, lipid from fatty acids and glycerol, protein from amino acids

✓ describe the tests for starch, glucose, lipid and protein

✓ describe a balanced diet, including carbohydrates, lipids, proteins, vitamins, minerals, water and dietary fibre

✓ recall the sources and functions of carbohydrates, lipids, proteins, vitamins A, C and D, and the mineral ions calcium and iron

✓ understand that energy requirements vary with activity levels, age and pregnancy

✓ understand variations in diet related to age, pregnancy, climate and occupation

✓ explain the dangers to health of protein deficiency and malnutrition

✓ recognise the structures of the human alimentary canal, and describe in outline the functions of the mouth, oesophagus, stomach, small intestine, large intestine and pancreas

✓ recall the types, structure and functions of teeth, understand the factors affecting their growth, and explain how to care for teeth and gums

✓ explain how and why food is moved through the gut by peristalsis, including the role of dietary fibre in the process

✓ understand the role of digestive enzymes, including the digestion of starch to glucose by amylase and maltase, the digestion of proteins to amino acids by proteases (pepsin, trypsin), and the digestion of lipids to fatty acids and glycerol by lipases

✓ recall that bile is produced by the liver and stored in the gall bladder, and understand the role of bile in neutralising stomach acid and emulsifying lipids

✓ understand how the structure of the villus helps absorption of the products of digestion in the small intestine

✓ describe how to carry out a simple experiment to determine the energy content in a food sample

✓ describe how to carry out a simple experiment with a digestive enzyme.

Questions

1 The diagram shows an experiment that was set up as a model to show why food needs to be digested.

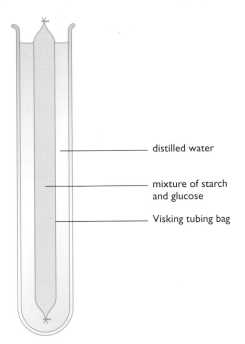

- distilled water
- mixture of starch and glucose
- Visking tubing bag

The Visking tubing acts as a model of the small intestine because it has tiny holes in it that some molecules can pass through. The tubing was left in the boiling tube for an hour, then the water in the tube was tested for starch and glucose.

a) Describe how you would test the water for starch, and for glucose. What would the results be for a 'positive' test in each case?

b) The tests showed that glucose was present in the water, but starch was not. Explain why.

c) If the tubing takes the place of the intestine, what part of the body does the water in the boiling tube represent?

d) What does 'digested' mean?

2 (Hint: page 4, Chapter 1 will help with this question.)
A student carried out an experiment to find out the best conditions for the enzyme pepsin to digest protein. For the protein, she used egg white powder, which forms a cloudy white suspension in water. The table opposite shows how the four tubes were set up.

Tube	Contents
A	5 cm³ egg white suspension, 2 cm³ pepsin, 3 drops of dilute acid. Tube kept at 37 °C
B	5 cm³ egg white suspension, 2 cm³ distilled water, 3 drops of dilute acid. Tube kept at 37 °C
C	5 cm³ egg white suspension, 2 cm³ pepsin, 3 drops of dilute acid. Tube kept at 20 °C
D	5 cm³ egg white suspension, 2 cm³ pepsin, 3 drops of dilute alkali. Tube kept at 37 °C

The tubes were left for 2 hours and the results were then observed. Tubes B, C and D were still cloudy. Tube A had gone clear.

a) Three tubes were kept at 37 °C. Why was this temperature chosen?

b) Explain what had happened to the protein in tube A.

c) Why did tube D stay cloudy?

d) Tube B is called a **control**. Explain what this means.

e) Tube C was left for another 3 hours. Gradually it started to clear. Explain why digestion of the protein happened more slowly in this tube.

f) The lining of the stomach secretes hydrochloric acid. Explain the function of this.

g) When the stomach contents pass into the duodenum, they are still acidic. How are they neutralised?

3 Copy and complete the following table of digestive enzymes.

Enzyme	Food on which it acts	Products
amylase		
trypsin		
		fatty acids and glycerol

4 Describe four adaptations of the small intestine (ileum) that allow it to absorb digested food efficiently.

5 Bread is made mainly of starch, protein and lipid. Imagine a piece of bread about to start its journey through the human gut. Describe what happens to the bread as it passes through the mouth, stomach, duodenum, ileum and colon. Explain how the bread is moved along the gut. Your description should be illustrated by two or three simplified diagrams. It would be easier to write up your account using a computer, leaving room for illustrations, or you might obtain these from websites or a CD-ROM.

6 The diagram shows a method that can be used to measure the energy content of some types of food. A student placed 20 cm³ of water in a boiling tube and measured its temperature. He weighed a small piece of pasta, and then held it in a Bunsen burner flame until it caught alight. He then used the burning pasta to heat the boiling tube of water, until the pasta had finished burning. Finally, he measured the temperature of the water at the end of the experiment.

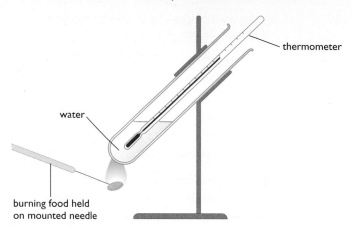

thermometer

water

burning food held
on mounted needle

To answer the questions that follow, use the following information.

- The density of water is 1 g/cm³.

- The pasta weighed 0.22 g.

- The water temperature at the start was 21 °C and at the end was 39 °C.

- The heat energy supplied to the water can be found from the formula:

energy (in joules) = mass of water × temperature change × 4.2

a) Calculate the energy supplied to the water in the boiling tube in joules (J). Convert this to kilojoules (kJ) by dividing by 1000.

b) Calculate the energy released from the pasta as kilojoules per gram of pasta (kJ/g).

c) The correct figure for the energy content of pasta is 14.5 kJ/g. The student's result is an underestimate. Write down three reasons why he may have got a lower than expected result. (Hint: think about how the design of the apparatus might introduce errors.)

d) Suggest one way the apparatus could be modified to reduce these errors.

e) The energy in a peanut was measured using the method described above. The peanut was found to contain about twice as much energy per gram as the pasta. Explain why this is the cas

Chapter 4: Blood and Circulation

The need for a circulatory system

Figure 4.1 shows the human circulatory system. Blood is pumped round and round a closed circuit made up of the heart and blood vessels. As it travels around, it collects materials from some places and unloads them at others. It transports many substances, including:

- oxygen from the lungs to all other parts of the body
- carbon dioxide from all parts of the body to the lungs
- nutrients from the gut to all parts of the body
- urea from the liver to the kidneys.

Hormones, antibodies and many other substances are also transported by the blood. It also distributes heat around the body.

Small animals, such as single-celled organisms, can get by without a circulatory system, relying on diffusion to exchange materials. A large animal, such as a human, needs a circulatory system for two main reasons:

- The human body has a small surface area to volume ratio.
- Distances in the human body are large.

Figure 4.2 shows how the ratio of surface area to volume of an animal decreases in bigger animals. In a very small organism such as the single-celled *Paramecium* (Figure 4.3), the surface area is large in proportion to the organism's volume. So substances like oxygen can be taken up by diffusion over the body surface at a fast enough rate to supply the organism's needs. In addition, the distances inside the *Paramecium's* body are small, so the oxygen doesn't have far to go.

We need a circulatory system to transport substances to and from the cells of the body. This chapter looks at the structure and function of the human circulatory system, the composition of blood, and medical problems associated with the heart and circulation.

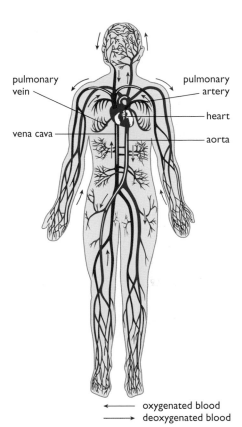

pulmonary vein

pulmonary artery

heart

vena cava

aorta

← oxygenated blood
→ deoxygenated blood

Figure 4.1 *The human circulatory system.*

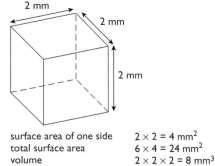

A small organism has a high surface area to volume ratio.

1 mm
1 mm
1 mm

surface area of one side	$1 \times 1 = 1$ mm^2
total surface area	$6 \times 1 = 6$ mm^2
volume	$1 \times 1 \times 1 = 1$ mm^3
surface area/volume	$6/1 = 6:1$

A larger organism has a lower surface area to volume ratio.

2 mm
2 mm
2 mm

surface area of one side	$2 \times 2 = 4$ mm^2
total surface area	$6 \times 4 = 24$ mm^2
volume	$2 \times 2 \times 2 = 8$ mm^3
surface area/volume	$24/8 = 3:1$

Figure 4.2 *An illustration of surface area to volume ratio. The bigger cube has a smaller surface area to volume ratio. It would be less able to obtain all the oxygen it needs through its surface.*

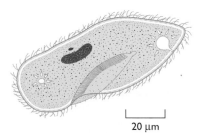

20 μm

Figure 4.3 *Unicellular organisms like this* Paramecium *do not have a circulatory system.*

Large animals such as humans cannot get all the oxygen they need through their surface (even if the body surface would allow it to pass through) – there just isn't enough surface to supply all that volume. To overcome this problem, humans have evolved special gas exchange organs (the lungs) and circulatory systems that transport materials in bulk around the body. The same idea applies to obtaining nutrients – the gut obtains nutrients from food and the circulatory system distributes the nutrients around the body.

Humans have a double circulatory system

One of the main functions of the circulatory system is to transport oxygen. Blood is pumped to the lungs to load oxygen. It is then pumped to the other parts of the body where it unloads the oxygen. In humans and other mammals, this is carried out by a **double circulation**. In a double circulation, the blood is pumped from the heart to the lungs. It then returns to the heart, before being pumped to the rest of the body (Figure 4.4).

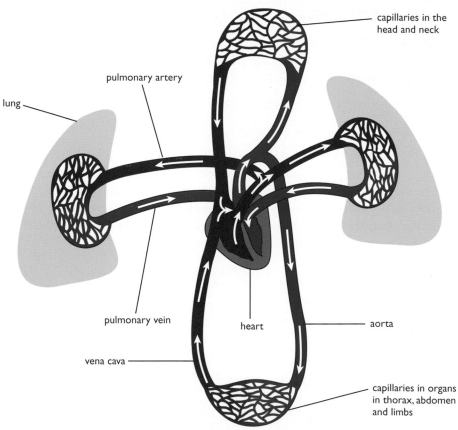

Figure 4.4 *The double circulation of blood in the human body. The blood passes through the heart twice during one complete circuit of the body.*

So there are two parts to a double circulation:

• the **pulmonary** circulation, in which blood is circulated to the lungs, where it is oxygenated

•the **systemic** circulation, in which blood is circulated through all the other parts of the body, where it unloads its oxygen.

Pulmonary means concerning the lungs.

A double circulation allows different blood pressures to be maintained in the pulmonary and systemic circulations. The pressure in the systemic circulation to the body is higher than the pressure in the pulmonary circulation to the lungs.

The human circulatory system comprises:

- the **heart** – this is a pump
- **blood vessels** – these carry the blood around the body; **arteries** carry blood away from the heart and towards other organs, **veins** carry blood towards the heart and away from other organs, **capillaries** carry blood through organs
- **blood** – the transport medium.

Figure 4.5 shows the main blood vessels in the human circulatory system.

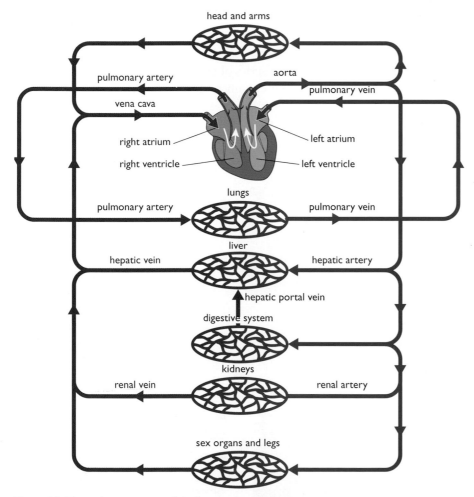

Figure 4.5 *The main components of the human circulatory system.*

The structure and function of the human heart

The human heart is a pump. It pumps blood around the body at different speeds and at different pressures according to the body's needs. It can do this because the wall of the heart is made from **cardiac muscle** (Figure 4.6).

Cardiac means 'to do with the heart'.

Cardiac muscle is unlike any other muscle in our bodies. It never gets fatigued ('tired') like skeletal muscle. On average, cardiac muscle fibres contract and then relax again about 70 times a minute. In a lifetime of 70 years, this special muscle will contract over two billion times – and never take a rest!

The bicuspid (mitral) and tricuspid valves are both sometimes called **atrioventricular** valves, as each controls the passage of blood from an atrium to a ventricle.

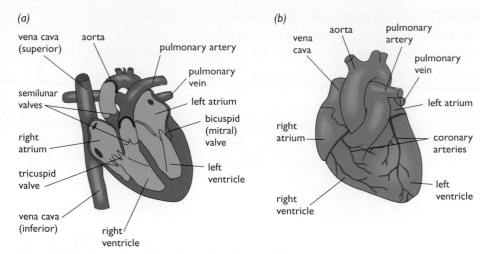

(a)

vena cava (superior)
aorta
pulmonary artery
pulmonary vein
semilunar valves
left atrium
bicuspid (mitral) valve
right atrium
tricuspid valve
left ventricle
vena cava (inferior)
right ventricle

(b)

aorta
pulmonary artery
vena cava
pulmonary vein
right atrium
left atrium
coronary arteries
right ventricle
left ventricle

Figure 4.6 *The human heart: (a) vertical section; (b) external view.*

Blood is moved through the heart by a series of contractions and relaxations of cardiac muscle in the walls of the four chambers. These events form the **cardiac cycle**. The main stages are illustrated in Figure 4.7.

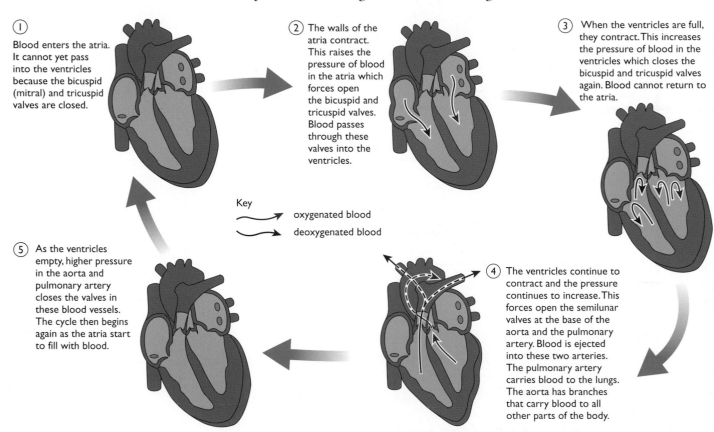

① Blood enters the atria. It cannot yet pass into the ventricles because the bicuspid (mitral) and tricuspid valves are closed.

② The walls of the atria contract. This raises the pressure of blood in the atria which forces open the bicuspid and tricuspid valves. Blood passes through these valves into the ventricles.

③ When the ventricles are full, they contract. This increases the pressure of blood in the ventricles which closes the bicuspid and tricuspid valves again. Blood cannot return to the atria.

⑤ As the ventricles empty, higher pressure in the aorta and pulmonary artery closes the valves in these blood vessels. The cycle then begins again as the atria start to fill with blood.

④ The ventricles continue to contract and the pressure continues to increase. This forces open the semilunar valves at the base of the aorta and the pulmonary artery. Blood is ejected into these two arteries. The pulmonary artery carries blood to the lungs. The aorta has branches that carry blood to all other parts of the body.

Key
→ oxygenated blood
→ deoxygenated blood

Figure 4.7 *The cardiac cycle.*

Atria is the plural of **atrium**.

When a chamber of the heart is contracting, we say it is in **systole**. When it is relaxing, we say it is in **diastole**.

The structure of the heart is adapted to its function in several ways.

- It is divided into a left side and a right side by the **septum**. The right ventricle pumps blood only to the lungs while the left ventricle pumps blood to all other parts of the body. This requires much more pressure, which is why the wall of the left ventricle is much thicker than that of the right ventricle.

- Valves ensure that blood can flow only in one direction through the heart.

- The walls of the atria are thin. They can be stretched to receive blood as it returns to the heart but can contract with enough force to push blood through the bicuspid and tricuspid valves into the ventricles.

- The walls of the heart are made of cardiac muscle which can contract and then relax continuously, without becoming fatigued.

- The cardiac muscle has its own blood supply – the **coronary circulation**. Blood reaches the muscle via **coronary arteries**. These carry blood to capillaries that supply the heart muscle with oxygen and nutrients. Blood is returned to the right atrium via **coronary veins**.

Heart rate

Normally the heart beats about 70 times a minute, but this can change according to circumstances. When we exercise, muscles must release more energy. They need an increased supply of oxygen for aerobic respiration (see Chapter 1). To deliver the extra oxygen, both the number of beats per minute (heart rate) and the volume of blood pumped with each beat (called stroke volume) increase.

When we are stressed (angry or afraid), our heart rate again increases. The increased output supplies extra blood to the muscles, enabling them to release extra energy through aerobic respiration. This allows us to fight or run away and is called the 'fight or flight' response. It is triggered by secretion of the hormone adrenaline from the adrenal glands (see Chapter 6).

When we sleep, our heart rate decreases as all our organs are working more slowly. They need to release less energy and so need less oxygen.

These changes in the heart rate are brought about by nerve impulses from a part of the brain called the **medulla** (Figure 4.8). When we start to exercise, our muscles produce more carbon dioxide in aerobic respiration. Sensors in the aorta and the carotid artery (the artery leading to the head) detect this increase. They send nerve impulses to the medulla. The medulla responds by sending nerve impulses along the accelerator nerve. When carbon dioxide production returns to normal, the medulla receives fewer impulses. It responds by sending nerve impulses along a decelerator nerve.

The accelerator nerve increases the heart rate. It also causes the heart to beat with more force and so increases blood pressure. The decelerator nerve decreases the heart rate. It also reduces the force of the contractions. Blood pressure then returns to normal.

These controls are both examples of **reflex actions** (see Chapter 6).

Butterflies in your stomach? Have you noticed a 'hollow' feeling in your stomach when you are anxious? There are no butterflies involved! It happens because blood that would normally flow to your stomach and intestines has been diverted to the muscles to allow the 'fight or flight' response.

The precise region of the medulla that controls heart functions is called the **cardiac centre**.

(not to scale)

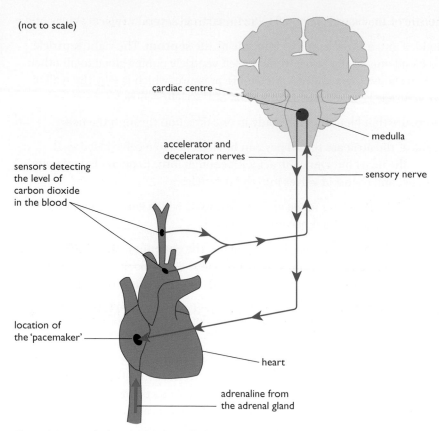

Figure 4.8 *How the heart rate is controlled.*

Arterioles are small arteries. They carry blood into organs from arteries. Their structure is similar to the larger arteries, but they have a larger proportion of muscle fibres in their walls. They are also **innervated** (have nerve endings in their walls) and so can be made to dilate (become wider) or constrict (become narrower) to allow more or less blood into the organ.

If *all* the arterioles constrict, it is harder for blood to pass through them – there is more resistance. This increases blood pressure. Prolonged stress can cause arterioles to constrict and so increase blood pressure.

All arteries carry **oxygenated blood** (blood containing a lot of oxygen) except the pulmonary artery, and the umbilical artery of an unborn baby. All veins carry **deoxygenated blood** (blood containing less oxygen) except the pulmonary vein and umbilical vein.

Arteries, veins and capillaries

Arteries carry blood from the heart to the organs of the body. This **arterial blood** is pumped out under high pressure by the ventricles of the heart. When blood leaves the heart through the aorta, valves in the aorta prevent the blood from returning to the heart, and the blood leaves in pressure 'spurts', stretching or distending the wall of the aorta. When the ventricle of the heart relaxes, the stretched section of the aorta recoils, increasing the pressure inside it. A wave of stretching followed by constriction passes along the aorta and through the arteries. This is the **pulse**. The pulse wave passes through the arteries and arterioles, getting weaker as it travels along, and disappears by the time it reaches the capillaries, so that there is no pulse in veins. Your pulse can be found anywhere an artery can be pressed against a bone, for example in the wrist.

To be able to stretch and recoil under the pressure wave, the walls of arteries have a thick layer of elastic tissue and smooth muscle fibres. The muscle in the walls of smaller arteries can contract to help the blood flow.

The pulse wave travels much faster than the actual blood flow within the arteries. By the time blood reaches the capillaries, it is only travelling at about 1 mm per second.

Veins carry blood from organs back to the heart. Blood flow through the capillaries reduces its pressure, so that blood in veins (**venous blood**) is at a very low pressure, much lower than that in the arteries. The walls of veins are thin, with little elastic fibres and smooth muscle. Figure 4.9 shows the structures of a typical artery and a typical vein with the same diameter.

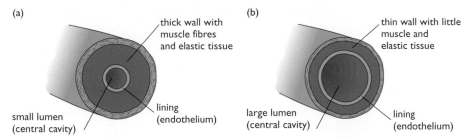

Figure 4.9 *The structure of (a) an artery and (b) a vein as seen in cross-section.*

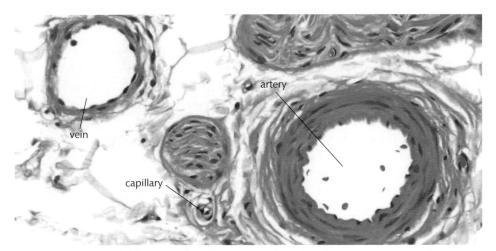

Figure 4.10 *Photograph of a section through an artery, vein and capillary.*

Veins also have valves called 'watch-pocket valves' which prevent the backflow of blood. The action of these valves is explained in Figure 4.11.

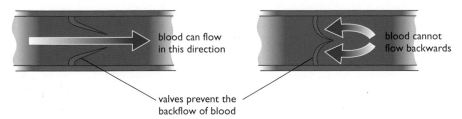

Figure 4.11 *The action of watch-pocket valves in veins.*

Capillaries carry blood through organs, bringing the blood close to every cell in the organ. Substances are transferred between the blood in the capillary and the cells. To do this, capillaries must be small enough to 'fit' between cells, and allow materials to pass through their walls easily. Figure 4.12 shows the structure of a capillary and how exchange of substances takes place between the capillary and nearby cells.

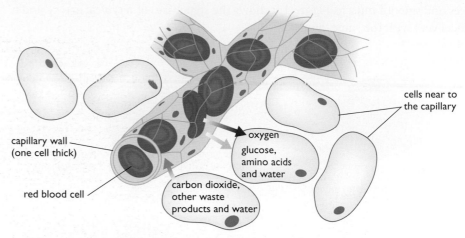

Figure 4.12 *How capillaries exchange materials with cells.*

Between the capillaries and the cells is a watery liquid called **tissue fluid**. The squamous epithelium cells of the capillary wall are leaky, so that the blood pressure causes fluid to leak out of the capillaries. Tissue fluid is similar in composition to blood plasma, except that it lacks proteins. These are too large to pass through the capillary cells. Tissue fluid forms a pathway for diffusion of substances between the capillaries and the cells. Most of the water from tissue fluid re-enters capillaries by osmosis, and some tissue fluid passes into another system of vessels called the **lymphatic system**. (Figure 4.13). The lymphatic system consists of vessels similar to blood capillaries, and are sometimes called lymphatic capillaries. They transport the fluid, called lymph, back to the blood (Figure 4.14). The lacteals of the small intestine are part of the lymphatic system (see Chapter 3, page 48).

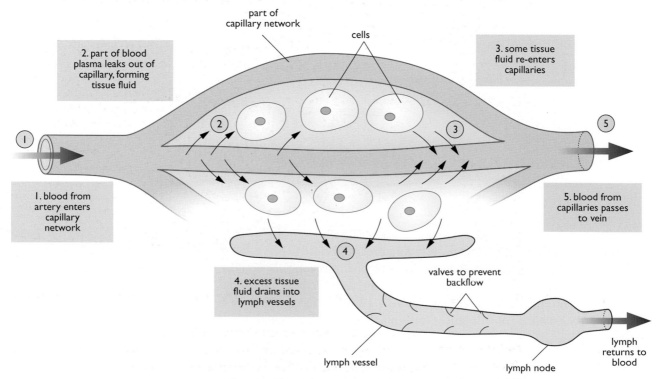

Figure 4.13 *Relationship between blood capillaries, tissue fluid and lymph.*

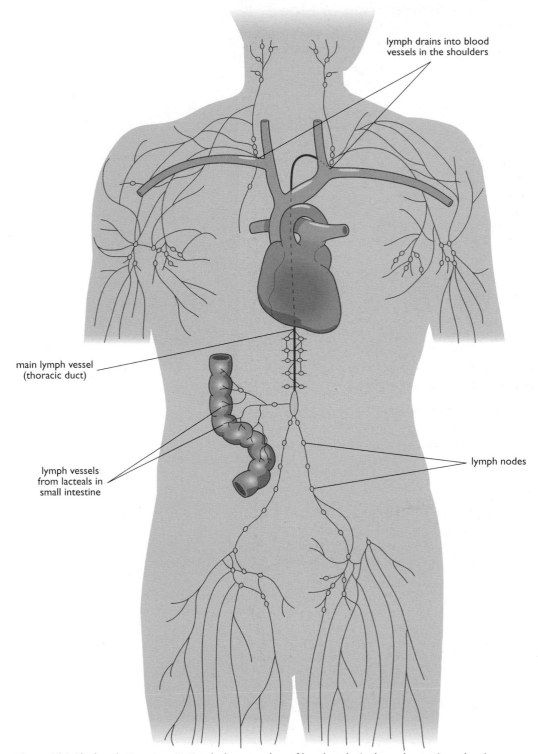

lymph drains into blood
vessels in the shoulders

main lymph vessel
(thoracic duct)

lymph vessels
from lacteals in
small intestine

lymph nodes

Figure 4.14 *The lymphatic system. Notice the large numbers of lymph nodes in the neck, armpits and groin. These may swell up if a person has an infection.*

Before lymph passes back into the blood it is filtered to remove dead cells and bacteria. This takes place in swellings called **lymph nodes**. Lymph nodes contain white blood cells that are important in destroying the harmful bacteria (see later in this chapter, page 63). When a person gets an infection, one of the first signs may be a swelling of the lymph nodes, often referred to as 'swollen glands'.

The composition of blood

Blood is a lot more than just a red liquid flowing through your arteries and veins! In fact, blood is a complex tissue. Figure 4.15 illustrates the main types of cells found in blood.

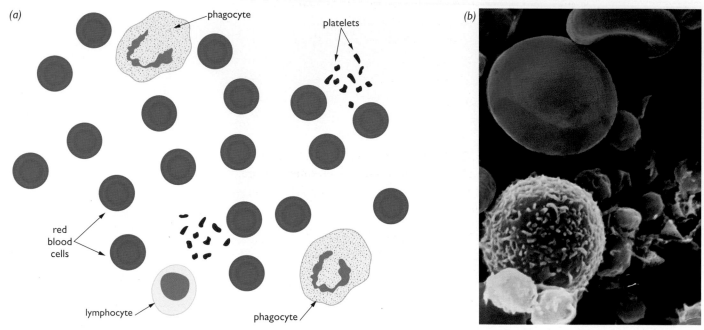

Figure 4.15 *The different types of blood cells (a) drawings of the different cells and (b) as seen in a photomicrograph.*

The different parts of blood have different functions. These are described in Table 4.1.

Component of blood	Description of component	Function of component
plasma	liquid part of blood: mainly water	carries the blood cells around the body; carries dissolved nutrients, hormones, carbon dioxide and urea; also distributes heat around the body
red blood cells (erythrocytes)	biconcave, disc-like cells with no nucleus; millions in each mm³ of blood	transport of oxygen – contain mainly haemoglobin, which loads oxygen in the lungs and unloads it in other regions of the body
white blood cells:		
lymphocytes	about the same size as red cells with a large spherical nucleus	produce antibodies to destroy microorganisms – some lymphocytes persist in our blood after infection and give us immunity to specific diseases
phagocytes	much larger cells with a large spherical or lobed nucleus	engulf bacteria and other microorganisms that have infected our bodies
platelets	the smallest cells – are really fragments of other cells	release chemicals to make blood clot when we cut ourselves

Table 4.1: *Functions of the different components of blood.*

Plasma

The liquid part of the blood is called plasma. It is a pale yellow, watery fluid containing many dissolved solutes. Plasma transports nutrients, such as glucose and amino acids, from the gut to the cells. It contains some soluble proteins, such as albumen and fibrinogen, as well as antibodies (immunoglobulins). Lipids are transported in the plasma attached to proteins (lipoproteins). The body's main nitrogenous waste product, urea, is carried in the plasma to the kidneys for excretion (Chapter 8). Hormones are also carried from the glands where they are made to the organs of the body where they have their effect (Chapter 6).

A small amount of carbon dioxide, about 10%, is carried inside red blood cells, attached to haemoglobin. However, most is transported in the plasma, either as dissolved carbon dioxide (5%) or as hydrogencarbonate ions, HCO_3^- (85%).

The blood plasma also transports heat around the body. More active organs, such as the muscles and liver, produce a lot of heat. This is transported to less metabolically active tissues, helping to maintain an even body temperature.

Red blood cells

The red blood cells or **erythrocytes** are highly specialised cells made in the bone marrow. They have a limited life span of about 100 days after which time they are destroyed in the spleen. They have only one function – to transport oxygen. Several features enable them to carry out this function very efficiently.

Red blood cells contain **haemoglobin**. This is an iron-containing protein that associates (combines) with oxygen to form **oxyhaemoglobin** when there is a high concentration of oxygen in the surroundings. We say that the red blood cell is loading oxygen. When the concentration of oxygen is low, oxyhaemoglobin turns back into haemoglobin and the red blood cell unloads its oxygen.

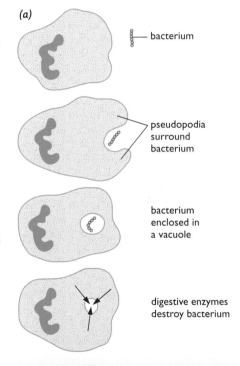

(a)

— bacterium

pseudopodia surround bacterium

bacterium enclosed in a vacuole

digestive enzymes destroy bacterium

$$\text{haemoglobin} + \text{oxygen} \underset{\text{low oxygen concentration (in tissues)}}{\overset{\text{high oxygen concentration (in lungs)}}{\rightleftharpoons}} \text{oxyhaemoglobin}$$

As red blood cells pass through the lungs, they load oxygen. As they pass through active tissues they unload oxygen.

Red blood cells do not contain a nucleus. It is lost during their development in the bone marrow. This means that more haemoglobin can be packed into each red blood cell so more oxygen can be transported. Their biconcave shape allows efficient exchange of oxygen in and out of the cell. Each red blood cell has a high surface area to volume ratio, giving a large area for diffusion. The thinness of the cell gives a short diffusion distance to the centre of the cell. Red blood cells are very flexible, allowing them to be squeezed through the blood capillaries, where the diameter of a capillary may be only just bigger than the diameter of a red blood cell.

White blood cells

There are several types of white blood cell. Their main role is to protect the body against invasion by disease-causing microorganisms (pathogens), such as bacteria and viruses. They do this in two main ways: **phagocytosis** and **antibody production**.

About 70% of white blood cells can ingest (take in) microorganisms such as bacteria. This is called phagocytosis, and the cells are **phagocytes**. They do this by changing their shape, producing extensions of their cytoplasm, called **pseudopodia**. The pseudopodia surround and enclose the microorganism in a vacuole. Once it is inside, the phagocyte secretes enzymes into the vacuole to break the microorganism down (Figure 4.16). Phagocytosis means 'cell eating' – you can see why it is called this.

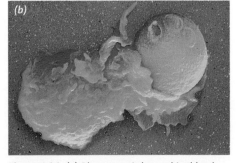

Figure 4.16 *(a) Phagocytosis by a white blood cell. (b) A phagocyte ingesting a yeast cell.*

Approximately 25% of white blood cells are **lymphocytes**. Their function is to make chemicals called **antibodies**. Antibodies are soluble proteins that pass into the plasma. Pathogens such as bacteria and viruses have telltale chemical 'markers' on their surfaces, which the antibodies recognise. These markers are called **antigens**. The antibodies stick to the surface antigens and destroy the pathogen. They do this in a number of ways, for example by:

- causing bacteria to stick together, so that phagocytes can ingest them more easily

- acting as a 'label' on the pathogen, so that it is more easily recognised by a phagocyte

- causing bacterial cells to burst open

- neutralising poisons (toxins) produced by pathogens.

Some lymphocytes do not get involved in killing microorganisms straight away. Instead, they develop into **memory cells**. Memory cells make us **immune** to a disease. These cells remain in the blood for many years, sometimes a lifetime. If the same microorganism re-infects, the memory lymphocytes start to reproduce and produce antibodies, so that the pathogen can be quickly dealt with (see Chapter 13).

Platelets

Platelets are not whole cells, but fragments of large cells made in the bone marrow. If the skin is cut, exposure to the air stimulates the platelets and damaged tissue to produce a chemical. This chemical causes the soluble plasma protein **fibrinogen** to change into insoluble fibres of another protein, **fibrin**. The fibrin forms a network across the wound, in which red blood cells become trapped. This forms a **clot**, which prevents further loss of blood and entry of pathogens. The clot develops into a scab, which protects the damaged tissue while new skin grows.

Antigens and organ transplants

All cells have surface antigens, and each of us has our own unique set of antigens on the cells of our bodies. Lymphocytes will not produce antibodies against these because they are 'self' antigens. The antigens on bacteria that infect us are 'non-self', so they are recognised as such by our lymphocytes, which produce antibodies to destroy them. The same is true of organs transplanted from other people. If a kidney, liver, heart or other organ is used for a transplant, the antigens on the cells of the organ will be recognised in the recipient's body as 'non-self'. Lymphocytes will produce antibodies to destroy these 'foreign' cells. This is called **organ rejection**.

Immunosuppressive drugs 'damp down' our immune responses. They reduce the risk of rejection of a transplanted organ, but they also reduce our ability to fight disease.

In looking for an donor organ, it is important to find one with antigens on the cells that match those of the patient as closely as possible. Because our antigens are determined by our genes, those of family members are often a good match. Those of identical twins are particularly closely matched. Finding an organ with antigens similar to the ones on the person needing the transplant is called **tissue typing**. A close match of antigens, together with the use of **immunosuppressive drugs**, considerably reduces the risk of rejection.

Blood groups and transfusions

Successful blood transfusions are only possible because of our knowledge of blood grouping (Figure 4.17). Blood grouping is a kind of tissue typing – blood of an inappropriate group will be 'rejected'.

Your blood group depends on the antigens on the surface of your red blood cells. Two of these are important in blood grouping. They are called the A and B antigens. The four possible blood groups are based on the presence of these antibodies and are called A, B, AB and O.

Besides the antigens on the red blood cells, each person also has antibodies in the plasma. These antibodies will destroy red blood cells with a particular antigen by making them **agglutinate** (clump together). Antibody 'a' agglutinates any red blood cells with the A antigen. Antibody 'b' agglutinates any red blood cells with the B antigen. Table 4.2 gives details of the different blood groups. As a general principle, remember that a person never carries the antibodies that would react with the antigens on their own red blood cells.

Blood group	Antigens on red cells	Antibodies in plasma
A	A	b
B	B	a
AB	AB	neither
O	neither	ab

Table 4.2: *Blood groups and their antigens and antibodies.*

It could be fatal for a doctor to give a blood transfusion of type A to a person with type B blood. The person receiving the blood has 'a' antibodies in their plasma that would make the red cells of the transfused blood agglutinate. Blood would clot inside the blood vessels and block them. The safety of a transfusion depends on the *antigens* on the red cells of the donated blood and the *antibodies* in the plasma of the person receiving the blood. If these can react (e.g. antigen A and antibody 'a'), then the transfusion will be unsafe. Table 4.3 shows safe and unsafe transfusions.

Blood group of donor		Blood group of recipient (antibodies present shown in brackets)			
Group	Antigen	A (b)	B (a)	AB (neither)	O (a +b)
A	A	✓	✗	✓	✗
B	B	✗	✓	✓	✗
AB	A + B	✗	✗	✓	✗
O	either	✓	✓	✓	✓

✓ = safe transfusion ✗ = unsafe transfusion

Table 4.3: *Blood transfusions and blood groups.*

Blood group O is sometimes called the universal donor because it can be given to any other blood group. Because there are no antigens on the surface of the red blood cells, there can be no reaction with any antibodies in the plasma of the person receiving the blood. Similarly, blood group AB is the universal recipient. Because there are no antibodies in the plasma, antigens in the donated blood cannot cause a reaction.

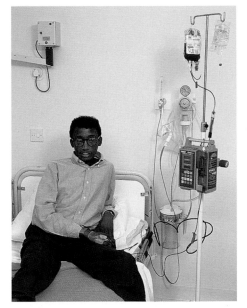

Figure 4.17 *A patient receiving a transfusion of blood.*

These antibodies differ from the antibodies we make against disease-causing organisms in one important way. They are present in our blood all the time, so we do not need to be exposed to the antigen to make the antibody.

Coronary heart disease

Coronary heart disease is caused by blockage of the coronary arteries, which supply blood to the heart muscle (Figure 4.18).

The coronary arteries are among the narrowest in the body. They are easily blocked by a build-up of fatty substances in their walls. If this happens, the oxygen supply to the heart muscle is reduced. The first symptoms of blockage may be a chest pain when the person exercises. This is called **angina**. The fatty deposits, called an **atheroma,** can build up to block the coronary artery completely, or may cause the formation of blood clots, which can also cause a blockage. If the artery is completely blocked, the oxygen supply to the heart muscle may be cut off altogether and the muscle stops contracting – a **heart attack** results.

If a large part of the heart muscle is affected, a heart attack is often fatal, although many heart attacks can be less severe, so that with treatment the person recovers.

A number of factors make coronary heart disease more likely.

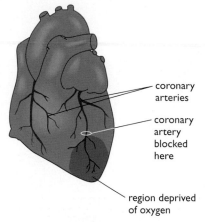

- Heredity – some people inherit genes that make them more liable to suffer from heart disease.

- High blood pressure – permanently high blood pressure is called **hypertension**. The heart has to work harder to pump the blood, which puts more strain on the heart muscle.

- Diet – a diet that is high in fat, especially saturated fat, causes raised blood cholesterol. High cholesterol in the blood combines with other lipids in forming an atheroma.

- Smoking – nicotine in tobacco smoke has several harmful effects. It constricts blood vessels, raising blood pressure, speeds up the heart rate, and increases blood cholesterol. Smoke also contains carbon monoxide, which lowers the ability of the blood to carry oxygen (see Chapter 2, page 29).

- Stress – hormones released during stress constrict blood vessels, raising blood pressure.

- Lack of exercise – regular exercise helps to reduce blood pressure and strengthens the heart.

Figure 4.18 *A blockage of a coronary artery cuts off the blood supply to part of the heart muscle.*

Exercise and the circulatory system

As soon as we start to exercise, our muscle cells respire more quickly to release more energy and produce more carbon dioxide. This triggers a reflex action to increase heart rate and blood pressure. More intense exercise results in a greater increase in heart rate and blood pressure.

During a period of exercise, the heart rate and blood pressure increase to a maximum to deliver the extra oxygen needed by the muscles. They remain at this level during the period of exercise, then begin to decrease to the pre-exercise levels as soon as the exercise ends. Figure 4.20 shows these changes.

Figure 4.19 *As we exercise, the circulatory system works harder.*

The period when heart rate and blood pressure are returning to normal following exercise is called the **recovery period**. They do not drop straight away to pre-exercise levels, but decrease gradually during this period. This is because, during exercise, lactic acid is formed in the muscles by anaerobic respiration (see Chapter 1). To get rid of lactic acid it must be oxidised, so as long as there is any lactic acid left in the muscles, extra oxygen will be needed to remove it. The heart must beat faster and with more force to deliver this extra oxygen. As the amount of lactic acid drops, so do the heart rate and blood pressure. When all the lactic acid has been oxidised, both return to pre-exercise levels. Figure 4.21 shows the changes in the heart rate and level of lactic acid in the blood following exercise. The length of recovery time is a rough measure of how healthy your heart is. If the heart can pump enough oxygen to the muscles during exercise, only a little lactic acid will be formed and the recovery period will be short.

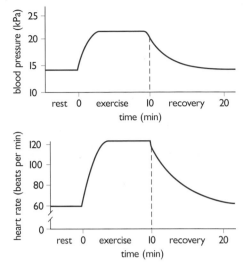

Figure 4.20 *How heart rate and blood pressure change during exercise.*

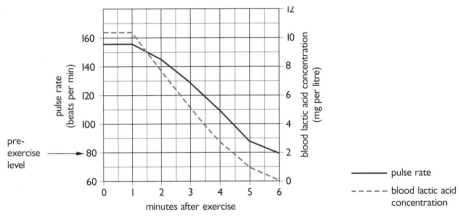

Figure 4.21 *The changes in pulse rate and lactic acid levels in the period following exercise.*

Regular exercise strengthens the heart muscle, and reduces the resting heart rate and blood pressure. Exercise doesn't just benefit the circulatory system, but improves a person's overall fitness in many ways. It:

- helps maintain a healthy body weight, preventing obesity

- reduces the levels of lipids, including cholesterol, in the blood

- builds skeletal muscle, increasing the mass and 'tone' of muscles (see Chapter 7)

- improves the strength of tendons and ligaments

- strengthens the diaphragm and intercostal muscles and increases the vital capacity of the lungs, so that breathing is more efficient

- stimulates the immune system

- helps maintain the level of glucose in the blood, reducing the risk of diabetes

- reduces the risk of contracting certain cancers, such as colon cancer

- makes people feel happier and more satisfied with life – some studies have even shown that regular exercise is as good as many medicines in treating depression.

End of Chapter Checklist

You should now be able to:

✓ recall the general plan of the circulation system, including the blood vessels to and from the heart, lungs, liver and kidneys

✓ describe the structure and function of the heart

✓ explain why the heart rate changes during exercise and under the influence of adrenaline

✓ recall the structure of arteries, veins and capillaries and understand their roles, including the nature of the pulse

✓ explain how tissue fluid is formed and recall its role

✓ recall the composition of the blood: red blood cells, white blood cells, platelets and plasma

✓ understand the role of plasma in the transport of carbon dioxide, digested food, urea, hormones and heat energy

✓ explain the adaptations of red blood cells for the transport of oxygen, including shape, structure and the presence of haemoglobin

✓ understand the role of white blood cells in preventing disease, by phagocytosis and antibody production

✓ understand the functions of blood clotting, and the role of platelets in this

✓ understand the problems associated with transplants, including the need to avoid rejection

✓ understand the nature of ABO blood groups and their importance in blood transfusions

✓ explain the causes of heart attacks

✓ understand the effects of exercise and the benefits of regular exercise on the circulatory system.

Questions

1 The circulation system carries nutrients, oxygen and carbon dioxide around the body.

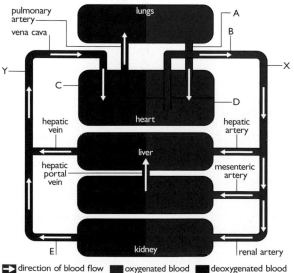

direction of blood flow ▣ oxygenated blood ▣ deoxygenated blood

a) Write down the correct labels for A to E.

b) Give two differences between the blood vessels at point X and point Y.

c) Which blood vessel contains the highest concentration of urea?

2 Blood transports oxygen and carbon dioxide around the body. Oxygen is transported by the red blood cells.

a) Give three ways in which a red blood cell is adapted to its function of transporting oxygen.

b) Describe how oxygen:

i) enters a red blood cell from the alveoli in the lungs

ii) passes from a red blood cell to an actively respiring muscle cell.

c) Describe how carbon dioxide is transported around the body.

3 Blood is carried around the body in arteries, veins and capillaries.

   a) Describe two ways in which the structure of an artery is adapted to its function.

   b) Describe three differences between arteries and veins.

   c) Describe two ways in which the structure of a capillary is adapted to its function.

4 The diagram shows a section through a human heart.

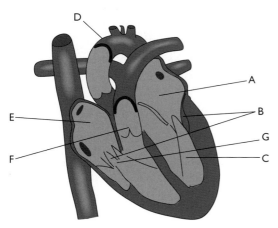

   a) Name the structures labelled A, B, C, D and E.

   b) What is the importance of the structures labelled B and F?

   c) Which letters represent the chambers of the heart to which blood returns:

   i)   from the lungs

   ii) from all the other organs of the body.

5 The diagram shows three types of cells found in human blood.

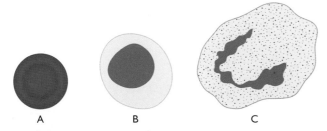

   a) Giving a reason for each answer, identify the blood cell which:

   i)   transports oxygen around the body

   ii) produces antibodies to destroy bacteria

   iii) engulfs and digests bacteria.

   b) Name one other component of blood found in the plasma and state its function.

6 The graph shows changes in a person's heart rate over a period of time.

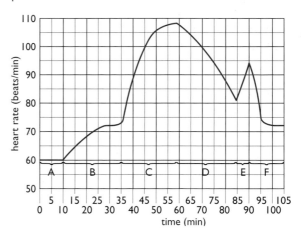

Giving reasons for your answers, give the letter of the time period when the person was:

   a) running

   b) frightened by a sudden loud noise

   c) sleeping

   d) waking.

7 The graph shows the changes that take place in heart rate before, during and after a period of exercise.

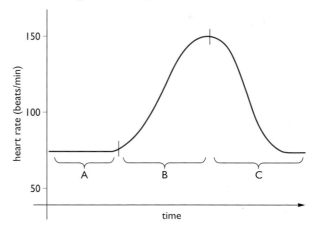

   a) Describe and explain the heart rates found:

   i)   at rest, before exercise (period A)

   ii) as the person commences the exercise (period B)

   iii) as the person recovers from the exercise (period C).

   b) How can the recovery period (period C) be used to assess a person's fitness?

   c) Write down four of the benefits of taking regular exercise.

8 *a)* Briefly describe how tissue fluid is formed.

   b) What is the function of tissue fluid?

Chapter 5: Coordination

Stimulus and response

Suppose you are walking along when you see a football coming at high speed towards your head. If your nerves are working properly, you will probably move or duck quickly to avoid contact. Imagine another situation where you are very hungry, and you smell food cooking. Your mouth might begin to 'water', in other words secrete saliva.

Each of these situations is an example of a **stimulus** and a **response**. A *stimulus* is a change in a person's surroundings, and a *response* is a reaction to that change. In the first example, the approaching ball was the stimulus, and your movement to avoid it hitting you was the response. The change in your environment was detected by your eyes, which are an example of a **receptor** organ. The response was brought about by contraction of muscles, which are an **effector** organ (they produce an effect). Linking the two is the nervous system, an example of a coordination system. A summary of the sequence of events is:

stimulus → receptor → coordination → effector → response

In the second example, the receptor for the smell of food was the nose, and the response was the secretion of saliva from glands. Glands secrete (release) chemical substances, and they are the second type of effector organ. Again, the link between the stimulus and the response is the nervous system. The information in the nerve cells is transmitted in the form of tiny electrical signals called nerve **impulses**.

Receptors

The role of any receptor is to detect the stimulus by changing its energy into the electrical energy of the nerve impulses. For example, the eye converts light energy into nerve impulses, and the ear converts sound energy into nerve impulses. When energy is changed from one form into another, this is called **transduction**. All receptors are **transducers** of energy (Table 5.1).

Receptor	Type of energy transduced
eye (retina)	light
ear (organ of hearing)	sound
ear (organ of balance)	mechanical (kinetic)
tongue (taste buds)	chemical
nose (organ of smell)	chemical
skin (touch/pressure/pain receptors)	mechanical (kinetic)
skin (temperature receptors)	heat
muscle (stretch receptors)	mechanical (kinetic)

Table 5.1: *Human receptors and the energy they transduce into electrical impulses.*

Notice how a 'sense' like touch is made up of several components. When we touch a warm surface we will be stimulating several types of receptor, including touch and temperature receptors, as well as stretch receptors in the muscles (see the section on skin later in this chapter). As well as this, each sense detects different aspects of the energy it receives. For example, the ears don't just detect sounds, but different loudness and frequencies of sound, while the eye not only forms an image, but also detects intensity of light and in humans can tell the difference between different light wavelengths (colours). Senses tell us a great deal about changes in our environment.

The central nervous system

The biological name for a nerve cell is a **neurone**. The impulses that travel along a neurone are not an electric current, as in a wire. They are caused by movements of charged particles (ions) in and out of the neurone. Impulses travel at speeds between about 10 and 100 m/s, which is much slower than an electric current, but fast enough to produce a rapid response.

Impulses from receptors pass along nerves containing **sensory neurones**, until they reach the **brain** and **spinal cord**. These two organs are together known as the **central nervous system**, or **CNS** (Figure 5.2).

Some animals can detect changes in their environment that are not sensed by humans. Insects such as bees can see ultraviolet (UV) light. The wavelengths of UV are invisible to humans (Figure 5.1).

(a)

(b)

Figure 5.1 *This yellow flower (a) looks very different to a bee, which sees patterns on the petals reflecting UV light (b).*

Figure 5.2 *The brain and spinal cord form the central nervous system. Cranial and spinal nerves lead to and from the CNS. The CNS sorts out information from the senses and sends messages to muscles.*

The CNS is well protected by the skeleton. The brain is encased in the skull or **cranium** (nerves connected to the brain are **cranial** nerves) and the spinal cord runs down the middle of the spinal column, passing through a hole in each vertebra. Nerves connected to the **spinal** cord are called spinal nerves.

Other nerves contain **motor neurones**, transmitting impulses to the muscles and glands. Some nerves contain only sensory or motor cells, while other nerves contain both – they are 'mixed'. A typical nerve contains thousands of individual neurones.

Both sensory and motor neurones can be very long. For instance, a motor neurone leading from the CNS to the muscles in the finger has a fibre about 1m in length, which is 100 000 times the length of the **cell body** (Figure 5.3).

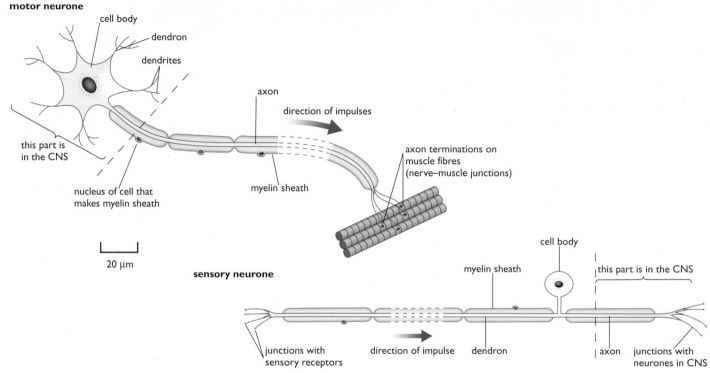

Figure 5.3 *The structure of motor and sensory neurones. The cell fibres (axon/dendron) are very long, which is indicated by the dashed sections.*

The cell body of a motor neurone is at one end of the fibre, in the CNS. The cell body has fine cytoplasmic extensions, called **dendrons**. These in turn form finer extensions, called **dendrites**. There can be junctions with other neurones on any part of the cell body, dendrons or dendrites. These junctions are called **synapses**. Later in this chapter we will deal with the importance of synapses in nerve pathways. One of the extensions from the motor neurone cell body is much longer than the other dendrons. This is the fibre that carries impulses to the effector organ, and is called the **axon**. At the end of the axon furthest from the cell body, it divides into many nerve endings. These fine branches of the axon connect with a muscle at a special sort of synapse called a **nerve–muscle junction**. In this way impulses are carried from the CNS out to the muscle. The signals from nerve impulses are transmitted across the nerve–muscle junction, causing the muscle fibres to contract. The axon is covered by a **sheath** made of a fatty material called **myelin**. The myelin sheath insulates the axon, preventing 'short circuits' with other axons, and also speeds up the conduction of the impulses. The sheath is formed by the membranes of special cells that wrap themselves around the axon as it develops.

A **sensory neurone** has a similar structure to the motor neurone, but the cell body is located on a side branch of the fibre, just outside the CNS. The fibre from the sensory receptor to the cell body is actually a dendron, while the fibre from the cell body to the CNS is a short axon. As with motor neurones, fibres of sensory neurones are often myelinated.

The eye

Many animals have eyes, but few show the complexity of the human eye. Simpler animals, such as snails, use their eyes to detect light but cannot form a proper image. Other animals, such as dogs, can form images but cannot distinguish colours. The human eye does all three. Of course it is not really the eye that 'sees' anything at all, but the brain that interprets the impulses from the eye. To find out how light from an object is converted into impulses representing an image, we need to look at the structure of this complex organ (Figure 5.4).

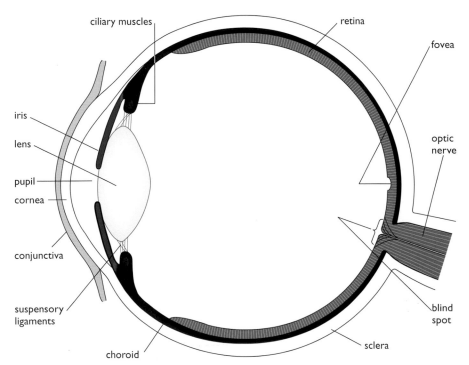

Figure 5.4 *A horizontal section through the human eye.*

The tough outer coat of the eye is called the **sclera**, which is the visible, white part of the eye. At the front of the eye the sclera becomes a transparent 'window' called the **cornea**, which lets light into the eye. Behind the cornea is the coloured ring of tissue called the **iris**. In the middle of the iris is a hole called the **pupil**, which lets the light through. It is black because there is no light escaping from the inside of the eye.

Underneath the sclera is a dark layer called the **choroid**. It is dark because it contains many pigment cells, as well as blood vessels. The pigment stops light being reflected around inside the eye. In the same way, the inside of a camera is painted matt black to stop stray light bouncing around and fogging the image on the film.

The fact that the inverted image is seen the right way up by the brain makes the point that it is the brain which 'sees' things, not the eye. An interesting experiment was carried out to test this. Volunteers were made to wear special inverting goggles for long periods. These turned the view of their surroundings upside down. At first this completely disorientated them, and they found it difficult to make even simple coordinated movements. However, after a while their brains adapted, until the view through the goggles looked normal. In fact, when the volunteers removed the goggles, the world then looked upside down!

The innermost layer of the back of the eye is the **retina**. This is the light-sensitive layer, the place where light energy is transduced into the electrical energy of nerve impulses. The retina contains cells called **rods** and **cones**. These cells react to light, producing impulses in sensory neurones. The sensory neurones then pass the impulses to the brain through the **optic nerve**. Rod cells work well in dim light, but they cannot distinguish between different colours, so the brain 'sees' an image produced by the rods in black and white. This is why we can't see colours very well in dim light: only our rods are working properly. The cones, on the other hand, will only work in bright light, and there are three types which respond to different wavelengths or colours of light – red, green and blue. We can see all the colours of visible light as a result of these three types of cones being stimulated to different degrees. For example, if red, green and blue are stimulated equally, we see white. Both rods and cones are found throughout the retina, but cones are particularly concentrated at the centre of the retina, in an area called the **fovea**. Cones give a sharper image than rods, which is why we can only see objects clearly if we are looking directly at them, so that the image falls on the fovea.

To form an image on the retina, light needs to be bent or **refracted**. Refraction takes place when light passes from one medium to another of a different density. In the eye, this happens first at the air/cornea boundary, and again at the lens (Figure 5.5). In fact the cornea acts as the first lens of the eye.

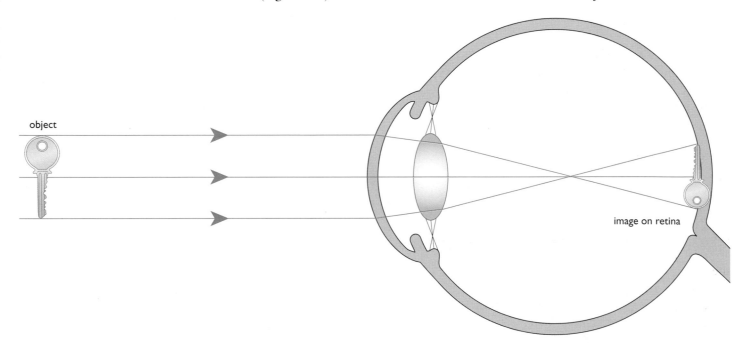

object

image on retina

Figure 5.5 *How the eye forms an image. Refraction of light occurs at the cornea and lens, producing an inverted image on the retina.*

As a result of refraction at the cornea and lens, the image on the retina is upside down, or **inverted.** The brain interprets the image the right way up. The role of the iris is to control the amount of light entering the eye, by changing the size of the pupil. The iris contains two types of muscles. **Circular muscles** form a ring shape in the iris, and **radial muscles** lie like the spokes of a wheel. In bright light, the pupil is made smaller, or **constricted**.

This happens because the circular muscles contract and the radial muscles relax. In dim light, the opposite happens. The radial muscles contract and the circular muscles relax, widening or **dilating** the pupil (Figure 5.6).

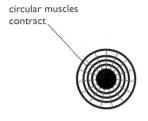

circular muscles contract

bright light
• circular muscles contract
• radial muscles relax
• pupil constricts

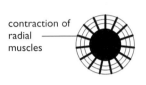

contraction of radial muscles

dim light
• circular muscles relax
• radial muscles contract
• pupil dilates

Figure 5.6 *The amount of light entering the eye is controlled by the iris, which alters the diameter of the pupil.*

Whenever our eyes look from a dim light to a bright one, the iris rapidly and automatically adjusts the pupil size. This is an example of a **reflex action**. You will find out more about reflexes later in this chapter. The purpose of the iris reflex is to allow the right intensity of light to fall on the retina. Light that is too bright could damage the rods and cones, and light that is too dim would not form an image. The intensity of light hitting the retina is the stimulus for this reflex. Impulses pass to the brain through the optic nerve, and straight back to the iris muscles, adjusting the diameter of the pupil. It all happens without the need for conscious thought – in fact we are not even aware of it happening.

There is one area of the retina where an image cannot be formed; this is where the optic nerve leaves the eye. At this position there are no rods or cones, so it is called the **blind spot**. The retina of each eye has a blind spot, but they are not a problem, because the brain puts the images from each eye together, cancelling out the blind spots of both eyes. As well as this, the optic nerve leaves the eye towards the edge of the retina, where vision is not very sharp anyway. To 'see' your own blind spot you can do a simple experiment. Cover or close your right eye. Hold this page about 30 cm from your eyes and look at the black dot below. Now, without moving the book or turning your head, read the numbers from left to right by moving your left eye slowly towards the right.

●   1   2   3   4   5   6   7   8   9   10   11   12   13   14   15

You should find that when the image of the dot falls on the blind spot it disappears. If you try doing this with both eyes open, the image of the dot will not disappear.

In the iris reflex, the route from stimulus to response is this:

stimulus (light intensity)
↓
retina (receptor)
↓
sensory neurones in optic nerve
↓
unconscious part of brain
↓
motor neurones in nerve to iris
↓
iris muscles (effector)
↓
response (change in size of pupil)

A way to prove to yourself that the eyes form two overlapping images is to try the 'sausage test'. Focus your eyes on a distant object. Place your two index fingers tip to tip, and bring them up in front of your eyes, about 30 cm from your face, while still focusing at a distance. You should see a finger 'sausage' between the two fingers. Now try this with one eye closed. What is the difference?

Why do we have two eyes?

Do you know which your 'dominant' eye is? Each eye forms a slightly different image of an object, as shown by the 'sausage test'. When your brain combines the images from each eye, it sees one image as dominant. You can find out which is your dominant eye by holding up a finger in line with a distant vertical object, such as a window frame. Now close each eye in turn. When you close the dominant eye, the finger will appear to move relative to the distant object. When you close the non-dominant eye, the finger will stay still.

There are several advantages to having two eyes. As well as cancelling the blind spot and providing a wider field of view, each eye forms a slightly different image of an object. The brain combines the information from each eye, giving us **stereoscopic** or three-dimensional (3D) vision, allowing us to judge the distance and depth of objects. This also allows us to estimate the speed of a moving object more accurately. You can show the benefit of stereoscopic vision by a simple experiment. Close one eye and ask a friend to hold a pencil horizontally in front of you, about 50 cm away. Try to line up your finger with the end of the pencil (without touching it). Now try the same thing with both eyes open. You will find it is harder to do using one eye.

Accommodation

The changes that take place in the eye which allow us to see objects at different distances are called **accommodation**.

You have probably seen the results of a camera or projector not being in focus – a blurred picture. In a camera, we can focus light from objects that are different distances away by moving the lens backwards or forwards, until the picture is sharp. In the eye, a different method is used. Rather than altering its position, the shape of the lens can be changed. A lens that is fatter in the middle (more convex) will refract light rays more than a thinner (less convex) lens. The lens in the eye can change shape because it is made of cells containing an elastic crystalline protein.

Figure 5.4 shows that the lens is held in place by a series of fibres called the **suspensory ligaments**. These are attached like the spokes of a wheel to a ring of muscle, called the **ciliary muscle**. The inside of the eye is filled with a transparent watery fluid which pushes outwards on the eye. In other words, there is a slight positive pressure within the eye. The changes to the eye that take place during accommodation are shown in Figure 5.7.

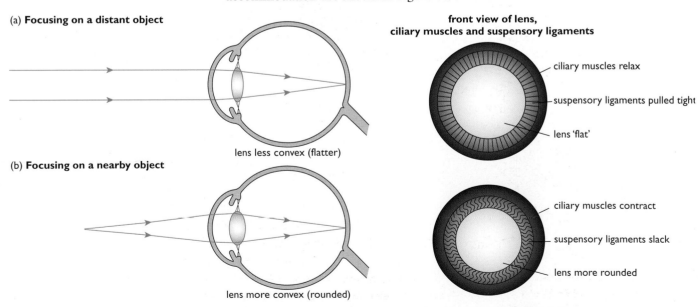

(a) **Focusing on a distant object**

lens less convex (flatter)

(b) **Focusing on a nearby object**

lens more convex (rounded)

front view of lens, ciliary muscles and suspensory ligaments

ciliary muscles relax

suspensory ligaments pulled tight

lens 'flat'

ciliary muscles contract

suspensory ligaments slack

lens more rounded

Figure 5.7 *Accommodation: how the eye focuses on objects at different distances.*

When the eye is focused on a distant object, the rays of light from the object are almost parallel when they reach the cornea (Figure 5.7a). The cornea refracts the rays, but the lens does not need to refract them much more to focus the light on the retina, so it does not need to be very convex. The ciliary muscles relax and the pressure in the eye pushes outwards on the lens, flattening it and stretching the suspensory ligaments. This is the condition when the eye is at rest – our eyes are focused for long distances.

When we focus on a nearby object, for example when reading a book, the light rays from the object are spreading out (diverging) when they enter the eye (Figure 5.7b). In this situation, the lens has to be more convex in order to refract the rays enough to focus them on the retina. The ciliary muscles now contract; the suspensory ligaments become slack and the elastic lens bulges outwards into a more convex shape.

The ear

Structure and function of the ear

The ear detects sound, and is also an organ of balance. The structure of the ear is shown in Figure 5.8.

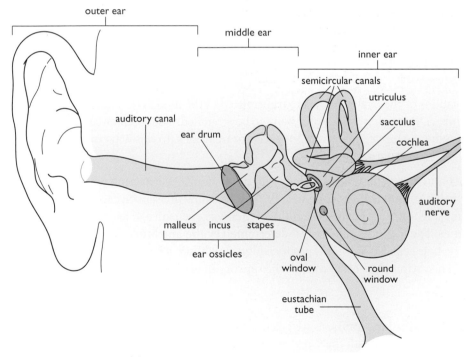

Figure 5.8 *The structure of the ear.*

The ear has three parts: the outer, middle and inner ear. The outer ear directs sound waves to the **ear drum** at the end of the auditory canal, causing it to vibrate. The vibrations are passed across the middle ear by three small bones (**ear ossicles**), the **malleus** (hammer), **incus** (anvil) and **stapes** (stirrup), which amplify the vibrations as they pass across them. The stapes transmits the vibrations to the **oval window** at one end of a coiled structure, the **cochlea**. The **eustachian tube** connects the middle ear with the throat, and allows the air pressure to be equalised either side of the ear drum.

Why do your ears 'pop' in an aeroplane? As an aeroplane gains height, the air pressure falls, causing the ear drum to bulge outwards. If you swallow, it opens the eustachian tube and allows air to pass into the middle ear from the throat, equalising the pressure. The 'pop' is caused by the eardrum going back to its normal position.

The receptor cells that convert the vibrations into nerve impulses are found in the cochlea, in a structure called the **organ of Corti**, which runs along the whole length of the coiled cochlea. To make this easier to understand, Figure 5.9 shows the cochlea 'uncoiled' and Figure 5.10 shows a cross-section of the cochlea.

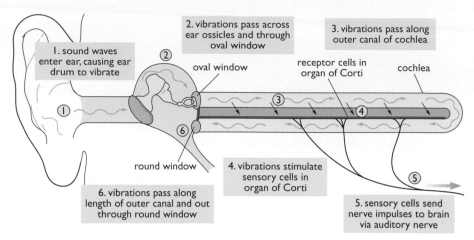

Figure 5.9 *Cochlea uncoiled to show the inner ear detects sounds.*

The outer fluid-filled canal runs from the oval window, all the way along the top of the cochlea, around the end of the cochlea and back to the **round window** at the end of the cochlea. Between the two parts of this canal is a middle chamber, also filled with fluid. The organ of Corti is located in a membrane between the middle and outer canals. The receptor cells in the organ of Corti have sensory 'hairs' embedded in a second membrane (Figure 5.10).

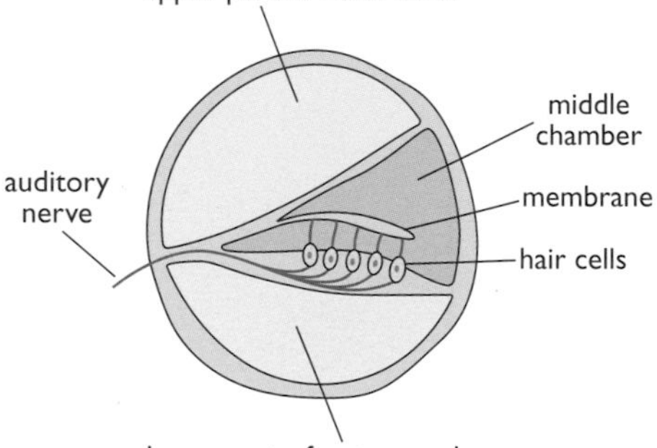

Vibrations of the oval window are transmitted to the fluid in the outer canal (Figure 5.9) causing the sensory hairs to be stretched. The receptor cells respond by producing nerve impulses in the receptor neurones. In this way sound is converted into nerve impulses.

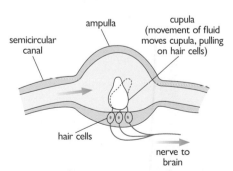

Figure 5.10 *Cross-section through the cochlea.*

The brain determines the frequency (pitch) of sounds by detecting which hair cells are being stimulated. Those nearest the oval window are sensitive to high-frequency sounds, while those nearest the round window are sensitive to low-frequency sounds. The loudness of sounds is determined by the amplitude (size) of vibrations of the hair cells. Loud sounds produce high-amplitude vibrations, which results in more nerve impulses per second in the sensory neurones.

Balance

The **semicircular canals**, as well as the **sacculus** and **utriculus**, are the parts of the ear involved with the sense of balance. In the swellings at the ends of the semicircular canals are more hair cells, with their hairs embedded in a jelly-like mass called a **cupula** (Figure 5.11).

Movement of fluid in the semicircular canals causes the cupula to pull on the hair cells, stimulating them to send nerve impulses to the brain. The canals are arranged in three planes at right-angles to each other, so that they can detect movement in any direction. The sacculus and utriculus also contain hair cells. Their hairs are embedded in a jelly containing calcium carbonate crystals, called an **otolith**. The weight of the otolith pulls on the hairs, stimulating the hair cells and producing nerve impulses. This gives information to the brain about the position of the head.

Figure 5.11 *Cross-section of an ampulla showing how movement is detected.*

Reflex actions

You saw on page 75 that the dilation and constriction of the pupil by the iris is an example of a reflex action. You now need to understand a little more about the nerves involved in a reflex. The nerve pathway of a reflex is called the **reflex arc**. The 'arc' part means that the pathway goes into the CNS and then straight back out again, in a sort of curve or arc (Figure 5.12).

A reflex action is a *rapid*, *automatic* (or *involuntary*) response to a stimulus. The action often (but not always) protects the body. Involuntary means that it is not started by impulses from the brain.

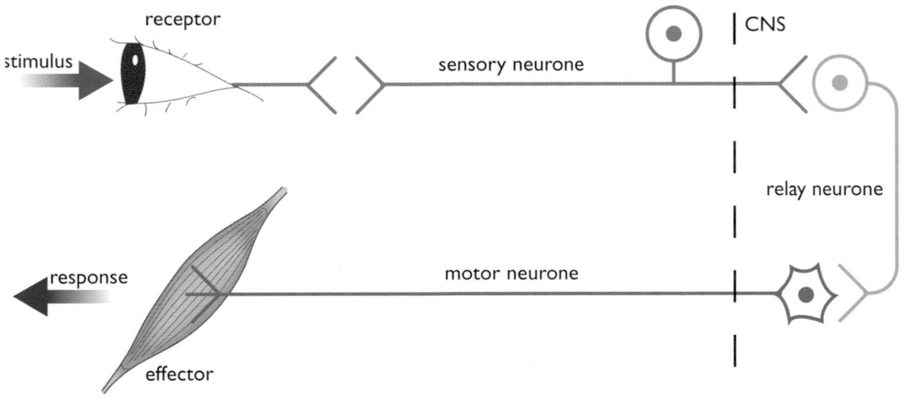

Figure 5.12 *Simplified diagram of a reflex arc.*

The iris–pupil reflex protects the eye against damage by bright light. Other reflexes are protective too, preventing serious harm to the body. Take, for example, the reflex response to a painful stimulus. This happens when part of your body, such as your hand, touches a sharp or hot object. The reflex results in your hand being quickly withdrawn. Figure 5.13 shows the nerve pathway of this reflex in more detail.

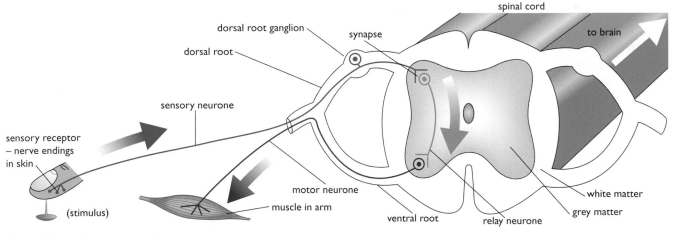

Figure 5.13 *A reflex arc in more detail.*

The stimulus is detected by temperature or pain receptors in the skin. These generate impulses in sensory neurones. The impulses enter the CNS through a part of the spinal nerve called the **dorsal root**. In the spinal cord the sensory neurones connect by synapses with short **relay neurones**, which in turn connect with motor neurones. The motor neurones emerge from the spinal cord through the **ventral root**, and send impulses back out to the muscles of the arm. These muscles then contract, pulling the arm (and thus finger) away from the harmful stimulus.

'Dorsal' and 'ventral' are words describing the back and front of the body. The dorsal roots of spinal nerves emerge from the spinal cord towards the back of the person, while the ventral roots emerge towards the front. Notice that the cell bodies of the sensory neurones are all located in a swelling in the dorsal root, called the **dorsal root ganglion**.

The middle part of the spinal cord consists mainly of nerve cell bodies, which gives it a grey colour. This is why it is known as **grey matter**. The outer part of the spinal cord is called **white matter**, and has a whiter appearance because it contains many axons with their fatty myelin sheaths.

Impulses travel through the reflex arc in a fraction of a second, so that the reflex action is very fast, and doesn't need to be started off by impulses from the brain. However, this doesn't mean that the brain is unaware of what is going on. This is because in the spinal cord, the reflex arc neurones also form synapses with nerve cells leading to and from the brain. The brain therefore receives information about the stimulus. This is how we feel the pain.

The knee-jerk reflex

You can demonstrate a spinal reflex on yourself quite easily. It is the well known **knee-jerk reflex**. Sit down and cross your legs, so that the upper leg hangs freely over the lower one. Grip the muscles of the top of the upper thigh with one hand and tap the area below the kneecap with a rubber hammer or the edge of the other hand (Figure 5.14). This may need a little practice, but you should eventually see the lower leg jerk forward as the muscles at the front of the thigh contract.

The reflex arc which brings this about is very similar to the withdrawal reflex (Figure 5.13), but in this case the stimulus is not detected in the skin, but in stretch receptors in the tendon below the knee. Tapping the tendon causes these receptors to be stretched. They react by sending nervous impulses towards the spinal cord through sensory neurones. The impulses then pass out again to the thigh muscles, through motor neurones, resulting in contraction of the muscle.

Of course, you would not normally experience a tap on the knee from a rubber hammer in everyday life. Where this reflex normally acts is in situations where the knee joint is unexpectedly flexed. For example, if you stumble, the stretch receptors will be stimulated in the same way, and the contraction of the thigh muscle will help to correct the stumble.

Movements are sometimes a result of reflex actions, but we can also contract our muscles as a **voluntary action**, using nerve cell pathways from the brain linked to the same motor neurones. A voluntary action is under *conscious control*.

Synapses

Synapses are critical to the working of the nervous system. The CNS is made of many billions of nerve cells, and these have links with many others, through synapses. In the brain, each neurone may form synapses with thousands of other neurones, so that there are an almost infinite number of possible pathways through the system.

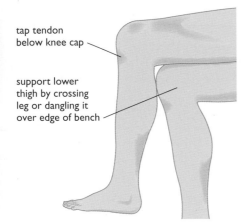

tap tendon below knee cap

support lower thigh by crossing leg or dangling it over edge of bench

Figure 5.14 *Demonstration of the knee-jerk reflex.*

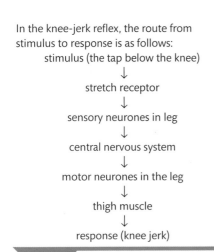

In the knee-jerk reflex, the route from stimulus to response is as follows:

stimulus (the tap below the knee)
↓
stretch receptor
↓
sensory neurones in leg
↓
central nervous system
↓
motor neurones in the leg
↓
thigh muscle
↓
response (knee jerk)

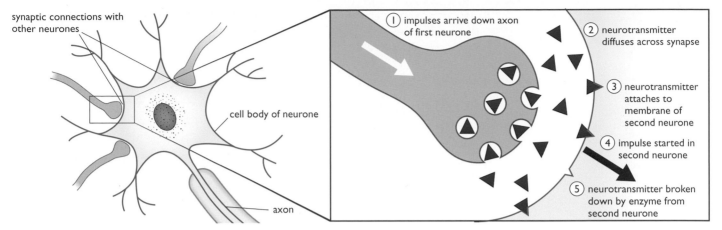

Figure 5.15 *The sequence of events happening at a synapse.*

A synapse is actually a *gap* between two nerve cells. The gap is not crossed by the electrical impulses passing through the neurones, but by chemicals. Impulses arriving at a synapse cause the ends of the fine branches of the axon to secrete a chemical, called a **neurotransmitter**. This chemical diffuses across the gap and attaches to the membrane of the second neurone. It then starts off impulses in the second cell (Figure 5.15). After the neurotransmitter has 'passed on the message', it is broken down by an enzyme.

Remember that many nerve cells, particularly those in the brain, have thousands of synapses with other neurones. The output of one cell may depend on the inputs from many cells adding together. In this way, synapses are important for integrating information in the CNS (Figure 5.16).

Because synapses are crossed by chemicals, it is easy for other chemicals to interfere with the working of the synapse. They may imitate the neurotransmitter, or block its action. This is the way that many well-known drugs, both useful and harmful, work.

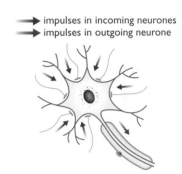

Figure 5.16 *Synapses allow the output of one nerve cell to be a result of integration of information from many other cells.*

The brain

The functions of different parts of the brain were first worked out through studies of people who had suffered brain damage through accident or disease. Nowadays we have very sophisticated electronic equipment that can record the activity in a normal living brain, but we are still relatively ignorant about the workings of this most complex organ of the body.

Your brain is sometimes called your 'grey matter'. This is because the positions of the grey and white matter are reversed in the brain compared with the spinal cord. The grey matter, mainly made of nerve cell bodies, is on the outside of the brain, and the axons that form the white matter are in the middle of the brain. The brain is made up of different parts, each with a particular function (Figure 5.17).

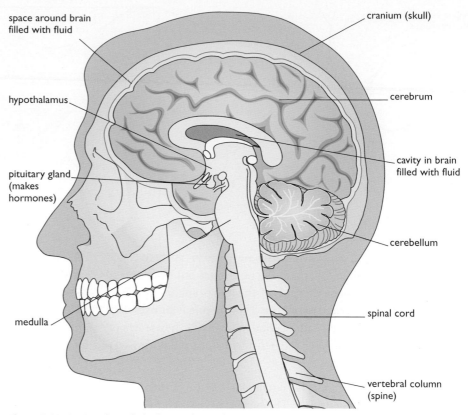

space around brain
filled with fluid

cranium (skull)

hypothalamus

cerebrum

pituitary gland
(makes
hormones)

cavity in brain
filled with fluid

cerebellum

medulla

spinal cord

vertebral column
(spine)

Figure 5.17 *Section through the human brain, showing its main parts.*

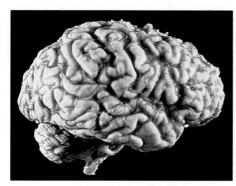

Figure 5.18 *A side view of a human brain. Notice the folded surface of the cerebral cortex.*

The largest part of the brain is the **cerebrum**, made of two **cerebral hemispheres**. The cerebrum is the source of all our conscious thoughts. It has an outer layer called the **cerebral cortex**, with many folds all over its surface (Figure 5.18).

The cerebrum has three main functions.

- It contains **sensory areas** that receive and process information from all our sense organs.

- It has **motor areas**, which are where all our voluntary actions originate.

- It is the origin of 'higher' activities, such as memory, reasoning, emotions and personality.

Different parts of the cerebrum carry out particular functions. For example, the sensory and motor areas are always situated in the same place in the cortex (Figure 5.19). Some parts of these areas deal with more information than others. Large parts of the sensory area deal with impulses from the fingers and lips, for example. This is illustrated in Figure 5.20.

Behind the cerebrum is the **cerebellum**. This region is concerned with coordinating the contraction of sets of muscles, as well as maintaining balance. This is important when you are carrying out complicated muscular activities, such as running or riding a bike. Underneath the cerebrum, connecting the spinal cord with the rest of the brain is the brain stem or **medulla**. This controls basic body activities such as heartbeat and breathing rate.

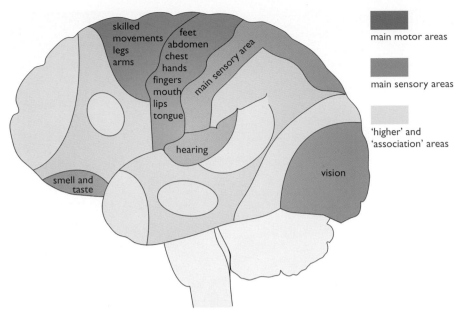

Figure 5.19 *Different parts of the cerebrum carry out specific functions.*

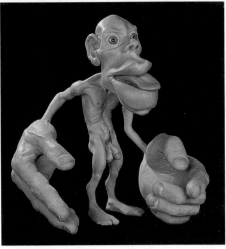

Figure 5.20 *A model of a human with its parts drawn in proportion to the amount of sensory information they send to the cortex of the brain (note that this does not apply to the eyes, which use more cortex than the rest of the body put together).*

The **pituitary gland** is located at the base of the brain, just below a part of the brain called the **hypothalamus**. The pituitary gland secretes a number of chemical 'messengers' called **hormones**, into the blood. The pituitary and hypothalamus are both discussed in Chapter 6.

Drugs and the nervous system

Drugs are chemicals that affect processes in a person's body. Many drugs are useful. For example, penicillin is an antibiotic that kills many of the bacteria that cause disease, and aspirin is an effective painkiller. However, a number of drugs act by interfering with the nervous system, and some of these can have very harmful side effects. A good example of this is the drug **nicotine** in tobacco. When a person smokes a cigarette, especially the first time, they get a 'buzz' from smoking. Their heart beats faster, their blood pressure rises and they feel excited. This is because nicotine is a **stimulant**, meaning that it increases brain activity. It does this by mimicking the action of neurotransmitters at the synapses of nerve cells in the brain. Some other stimulants also affect synapses. **Caffeine**, the mild drug in tea and coffee, causes more neurotransmitter than normal to be released.

The opposite of a stimulant is a **depressant** drug. One example is **alcohol** (ethanol). The alcohol in beer, wine and spirits slows down the nervous system, even when drunk in small quantities, and increases the time a person takes to react to a stimulus. That is why driving after drinking alcohol is so dangerous. The driver will not react quickly to sudden danger, such as a person walking into the road (Figure 5.21).

Larger amounts of alcohol in the body interfere with the drinker's balance and muscular control, and lead to blurred vision and slurred speech. High concentrations of alcohol in the blood can even cause coma and death.

Figure 5.21 *Alcohol in the bloodstream increases reaction times and is a common cause of car accidents.*

Many people drink moderate amounts of alcohol to relax. However, to some people alcohol is an addictive drug, and long-term alcohol abuse leads to serious medical problems. Alcohol is quickly absorbed into the blood through the stomach and intestines, and is taken around the body. The liver breaks the alcohol down (**detoxification**), but if a person drinks large amounts regularly it may not be able to cope. The person can get a disease called **cirrhosis** of the liver, where the liver does not perform its functions properly and toxins in the blood build up to high levels. This disease is usually fatal. Alcohol also damages the brain and stomach lining.

You should now be able to:

✓ recall the basic plan of the central nervous system

✓ recall that there are receptors which respond to heat, chemical, mechanical, sound and light energy

✓ recall the structures and functions of motor, sensory and relay neurones

✓ understand the initiation of the nerve impulse in receptors and direction of movement along a neurone

✓ recall the structure and function of the eye as a receptor

✓ explain the function of the eye in focusing on near and distant objects, and in responding to changes in light intensity

✓ understand the functions of stereoscopic vision

✓ recall the structure of the ear, and describe its functions in balance and hearing

✓ describe the structure of the reflex arc

✓ recall spinal reflexes, including the withdrawal reflex and the knee-jerk reflex

✓ understand the transfer of impulses across a synapse

✓ recall the main areas of the brain and an outline of their functions, including the cerebral hemispheres, cerebellum, medulla, pituitary gland and hypothalamus

✓ describe the effects of excess alcohol on the nervous system and behaviour

Questions

1 A cataract is an eye problem suffered by some people, especially the elderly. The lens of the eye becomes opaque (cloudy) which blocks the passage of light. It can lead to blindness. Cataracts can be treated by a simple eye operation, where a surgeon removes the lens. After the operation, the patient is able to see again, but the eye is unable to carry out accommodation, and the patient will probably need to wear glasses.

   a) Explain why the eye can still form an image after the lens has been removed.

   b) What is meant by 'accommodation'? Why is this not possible after a cataract operation?

   c) Will the patient need glasses to see nearby or distant objects clearly? Explain your answer.

2 The diagram shows a section through a human eye.

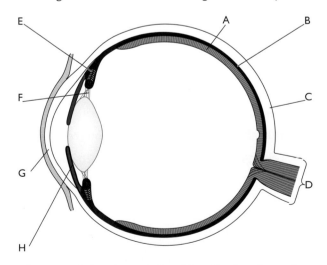

   a) The table (on the next page) lists the functions of some of parts A to H. Copy the table and write the letters of the correct parts in the boxes.

Function	Letter
refracts light rays	
converts light into nerve impulses	
contains pigment to stop internal reflection	
contracts to change the shape of the lens	
takes nerve impulses to the brain	

b) i) Which label shows the iris?

   ii) Explain how the iris controls the amount of light entering the eye.

   iii) Why is this important?

3 The diagram shows some parts of the nervous system involved in a simple reflex action that happens when a finger touches a hot object.

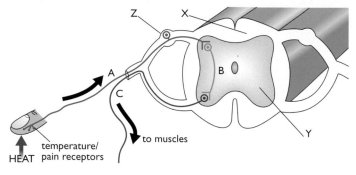

a) What type of neurone is:

   i) neurone A   ii) neurone B   iii) neurone C?

b) Describe the function of each of these types of neurone.

c) Which parts of the nervous system are shown by the labels X, Y and Z?

d) In what form is information passed along neurones?

e) Explain how information passes from one neurone to another.

4 **a)** Which part of the human brain is responsible for controlling each of the following actions:

   i) keeping your balance when you walk

   ii) maintaining your breathing when you are asleep

   iii) making your leg muscles contract when you kick a ball?

b) A 'stroke' is caused by a blood clot blocking the blood supply to part of the brain.

   i) One patient, after suffering a stroke, was unable to move his left arm. Which part of his brain was affected?

   ii) Another patient lost her sense of smell following a stroke. Which part of her brain was affected?

5 **a)** List five examples of stimuli that affect the body and state the response produced by each stimulus.

b) For one of your five examples, explain:

   i) the nature and role of the receptor

   ii) the nature and role of the effector organ.

c) For the same example, describe the chain of events from stimulus to response.

6 The diagram shows the parts of the ear.

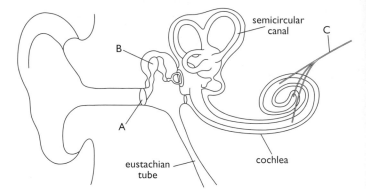

a) Name the parts labelled A, B and C.

b Explain the role of the ear bones (ear ossicles).

c) Explain how vibrations in the fluid inside the cochlea are converted into nerve impulses.

d) What is the function of the eustachian tube?

e) How does the ear distinguish between:

   i) sounds of different frequencies (pitch)

   ii) sounds of different loudness?

Chapter 6: Chemical Coordination

Glands and hormones

A gland is an organ that releases or **secretes** a substance. This means that cells in the gland make a chemical which passes out of the cells. The chemical then travels somewhere else in the body, where it carries out its function. There are two types of glands – **exocrine** and **endocrine** glands. Exocrine glands secrete their products through a tube or **duct**. For example, salivary glands in your mouth secrete saliva down salivary ducts, and tear glands secrete tears through ducts that lead to the surface of the eye. Endocrine glands have no duct, and so are called **ductless** glands. Instead, their products, the hormones, are secreted into the blood vessels that pass through the gland (Figure 6.1).

This chapter looks at some of the main endocrine glands and the functions of the hormones they produce. Because hormones are carried in the blood, they can travel to all areas of the body. They usually only affect certain tissues or organs, called 'target organs', which can be a long distance from the gland that made the hormone. Hormones only affect particular tissues or organs if the cells of that tissue or organ have special chemical receptors for the particular hormone. For example, the hormone insulin affects the cells of the liver, which have insulin receptors.

The differences between nervous and endocrine control

Although the nervous and endocrine systems both act to coordinate body functions, there are differences in the way that they do this. These are summarised in Table 6.1.

Nervous system	Endocrine system
• works by nerve impulses transmitted through nerve cells (although chemicals are used at synapses)	• works by hormones transmitted through the bloodstream
• nerve impulses travel fast and usually have an 'instant' effect	• hormones travel more slowly and generally take longer to act
• response is usually short-lived	• response is usually longer-lasting
• impulses act on individual cells such as muscle fibres, so have a very localised effect	• hormones can have widespread effects on different organs (although they only act on particular tissues or organs if the cells have the correct receptors)

Table 6.1: *The nervous and endocrine systems compared.*

> *The nervous system (Chapter 5) is a coordination system forming a link between stimulus and response. The body has a second coordination system, which does not involve nerves. This is the **endocrine** system. It consists of organs called endocrine **glands**, which make chemical messenger substances called **hormones**. Hormones are carried in the bloodstream.*

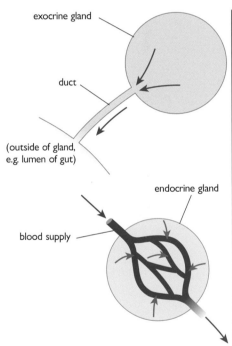

exocrine gland

duct

(outside of gland, e.g. lumen of gut)

endocrine gland

blood supply

Figure 6.1 *Exocrine glands secrete their products though a duct, while endocrine glands secrete hormones into the blood.*

> The receptors for some hormones are located in the cell membrane of the target cell. Other hormones have receptors in the cytoplasm, and some in the nucleus. Without specific receptors, a cell will not respond to a hormone at all.

The positions of the endocrine glands

The main endocrine glands are shown in Figure 6.2. Table 6.2 lists some of the hormones that they make, and their functions. Several of these hormones will be covered in other chapters of this book, where relevant.

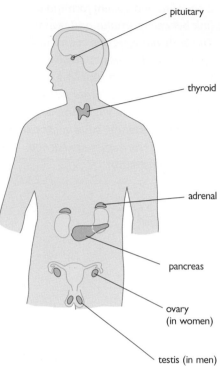

Figure 6.2 *The main endocrine glands of the body.*

Gland	Hormone	Some functions of the hormones
pituitary	follicle stimulating hormone (FSH)	stimulates egg development and oestrogen secretion in females and sperm production in males
	luteinising hormone (LH)	stimulates egg release (ovulation) in females and testosterone production in males
	anti-diuretic hormone (ADH)	controls the water content of the blood (see Chapter 8)
	growth hormone (GH)	speeds up the rate of growth and development in children
thyroid	thyroxin	controls the body's metabolic rate (how fast chemical reactions take place in cells)
pancreas	insulin	lowers blood glucose
	glucagon	raises blood glucose
adrenals	adrenaline	prepares body for physical activity
testes	testosterone	controls development of male secondary sexual characteristics
ovaries	oestrogen	controls development of female secondary sexual characteristics
	progesterone	regulates menstrual cycle

Table 6.2: *Some of the main endocrine glands, the hormones they produce and their functions.*

The **pituitary** gland is found at the base of the brain. It secretes a number of hormones, including several that regulate reproduction. These are dealt with in Chapter 9. The pituitary contains neurones linking it to a part of the brain called the **hypothalamus**, and some of its hormones are produced under the control of the brain. For example, **antidiuretic hormone** (ADH) is made in the cell bodies of neurones in the hypothalamus. It is secreted from the ends of fibres of these neurones and stored in the pituitary gland. The target organs of ADH are the kidneys, where the hormone is involved in controlling the water content of the blood. How this happens is described in much more detail in Chapter 8, but a key point is that the water potential of the blood is monitored by receptors in the hypothalamus, in turn controlling ADH release by the pituitary. This shows how the nervous system and endocrine system can sometimes act together in coordinating a response.

The **thyroid** is a gland in the neck, shaped a little like a butterfly. It secretes several hormones, including thyroxine, which speeds up the metabolic rate of the body. The metabolic rate is the rate at which the cells of the body carry out all their chemical reactions. The release of thyroxine is also under the control of the hypothalamus of the brain (see below).

The **pancreas** is both an endocrine and an exocrine organ. It secretes two hormones called **insulin** and **glucagon**, which are both involved in the regulation of blood glucose. The pancreas is also a gland of the digestive system, secreting enzymes through the pancreatic duct into the small intestine (see Chapter 3). The endocrine functions of the pancreas are described later in this chapter.

The sex organs of males and females, as well as producing sex cells (gametes) are also endocrine organs. Both the **testes** and **ovaries** make hormones involved in the control of reproduction. This topic is covered in Chapter 9.

We will now look at the functions of three endocrine glands in more detail: the thyroid, adrenals and pancreas.

The thyroid and control of metabolic rate

The thyroid makes a number of hormones, including thyroxine. The thyroid gland is the only endocrine gland that stores large amounts of a hormone. All other glands make their hormones as required, but the thyroid contains about a 100-day supply of thyroxine. Thyroxine contains the element iodine. Lack of iodine in the diet is one of the causes of an enlarged thyroid gland, called a goitre (Figure 6.3).

If a person fasts (eats nothing) overnight, and their rate of oxygen consumption at rest is then measured, this is called the basal metabolic rate (BMR). Thyroxine increases the BMR, stimulating cells to respire aerobically to produce more ATP (see Chapter 1, page 7).

A rise in metabolic rate is needed by the body in various situations, such as during exercise and in cold surroundings, when ATP can be broken down to yield heat to maintain body temperature. Thyroxine also increases protein synthesis in the body, so it is needed for normal growth and development in children.

However, when changes are not needed, the BMR is controlled within very strict limits. The hypothalamus in the brain detects the slightest drop in the metabolic rate, and responds by making a hormone called **thyrotropin releasing hormone** (TRH). TRH travels the short distance from the hypothalamus to the pituitary, where it stimulates the release of yet another hormone, called **thyroid stimulating hormone** (TSH). TSH passes in the bloodstream to the thyroid, where it stimulates synthesis and release of thyroxine.

Thyroxine causes the metabolic rate to return to normal. In turn this 'switches off' production of TRH, TSH and thyroxine release (Figure 6.4).

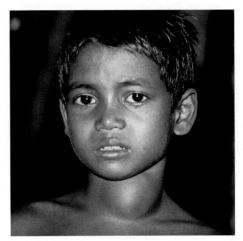

Figure 6.3 *An enlarged thyroid gland, called a goitre.*

Negative feedback

Control of metabolic rate is a good example of a principle called **negative feedback**. Negative feedback is a way that control is exercised in the body. During negative feedback, a change in conditions in the body is detected, starting a process that works to return conditions to normal. Not just metabolic rate, but many other conditions are monitored and maintained by negative feedback, such as body temperature (see Chapter 8, page 117). Many negative feedback processes involve hormones as 'messengers', such as ADH in the control of the water content of the blood (Chapter 8, page 112). The information pathway forms a loop, so it is also known as a **negative feedback loop**. The negative feedback loop controlling metabolic rate is shown in Figure 6.4.

Negative feedback depends on there being a normal, 'correct' value for the condition that is being monitored. For example, our normal body temperature is 37 °C. Biologists call the normal value the **set-point**. The set-point is monitored by the body, and any deviation from the set-point is corrected by negative feedback.

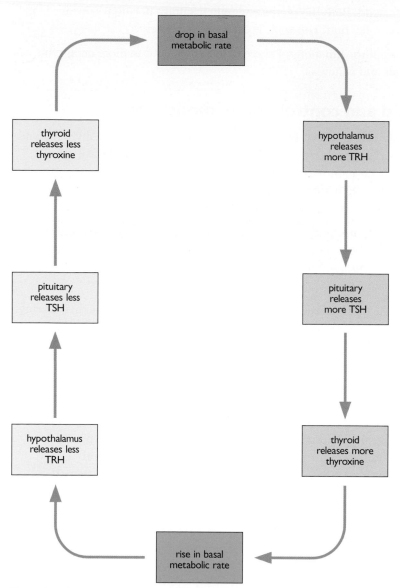

Figure 6.4 *Thyroxine controls metabolic rate by a negative feedback loop.*

Adrenaline – the 'fight or flight' hormone

'Adrenal' means 'next to the kidneys' which describes where the adrenal glands are located, on top of these organs (see Figure 6.2).

When you are frightened, excited or angry, your **adrenal** glands secrete the hormone **adrenaline**.

Adrenaline acts at a number of target organs and tissues, preparing the body for action. In animals other than humans this action usually means dealing with an attack by an enemy, where the animal can stay and fight or run away – hence 'fight or flight'. This is not often a problem with humans, but there are plenty of other times when adrenaline is released.

Figure 6.5 *Many human activities cause adrenaline to be produced, not just a 'fight or flight' situation!*

When the body prepares for action, its muscles will need a good supply of oxygen and glucose for respiration. Adrenaline causes several changes to take place to ensure this happens, along with other changes to prepare for 'fight or flight'.

- The breathing rate increases and breaths become deeper, taking more oxygen into the body.

- The heart beats faster, sending more blood to the muscles, so that they receive more glucose and oxygen for respiration.

- Blood is diverted away from the intestine and into the muscles.

- In the liver, stored carbohydrate is changed into glucose and released into the blood. The muscle cells absorb more glucose and use it for respiration.

- The pupils dilate, increasing visual sensitivity to movement.

- Mental awareness is increased, so reactions are faster.

In humans, adrenaline is not just released in a 'fight or flight' situation, but in many other stressful activities too, such as preparing for a race, going for a job interview or taking an exam.

Controlling blood glucose

You saw earlier that adrenaline can raise blood glucose from stores in the liver. The liver cells contain carbohydrate in the form of **glycogen**. Glycogen is made from long chains of glucose subunits joined together (see Chapter 3), producing a large, insoluble molecule. Being insoluble makes glycogen a good storage product. When the body is short of glucose, the glycogen can be broken down into glucose, which then passes into the bloodstream.

Adrenaline raises blood glucose concentration in an emergency, but two other hormones act all the time to control the level, keeping it fairly constant. Both these hormones are made by the pancreas.

Insulin stimulates removal of glucose from the bloodstream into cells and causes the liver cells to convert glucose into glycogen. This lowers the glucose concentration in the blood when it is too high.

The other hormone is **glucagon**. This stimulates the liver cells to break down glycogen into glucose, raising the concentration of glucose in the blood if it is too low.

Together, they work to keep the blood glucose approximately constant, at a little less than 1 g of glucose in every dm^3 of blood. Both hormones are released by special cells in the pancreas, in direct response to the level of glucose in the blood passing through this organ. In other words:

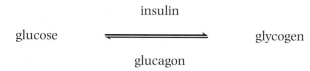

The concentration of glucose in your blood will start to rise after you have had a meal. Sugars from digested carbohydrate pass into the blood and are carried to the liver in the hepatic portal vein (Chapter 4). Here the glucose is converted to glycogen, so the blood leaving the liver in the hepatic vein will have a lower concentration of glucose.

Control of blood glucose levels by both insulin and glucagon operates through negative feedback systems. If the glucose concentration of the blood rises, release of insulin brings about a drop in glucose levels, until the normal level of glucose is achieved again. Glucagon works the other way – if glucose levels fall, glucagon stimulates responses that raise the glucose level. Having two hormones with opposite effects allows for better control.

Diabetes

Some people have a disease where their pancreas cannot make enough insulin to keep their blood glucose level constant – it rises to very high concentrations. The disease is called **diabetes**, or sometimes 'sugar diabetes'. One symptom of diabetes can be detected by a chemical test on urine. Normally, people have no glucose at all in their urine. Someone suffering from diabetes may have such a high concentration of glucose in the blood that it is excreted in their urine. This can be shown up by using coloured test strips (Figure 6.7).

Most of the cells of the pancreas are concerned with making digestive enzymes. However, in the pancreas tissue, there are small groups of cells called the **Islets of Langerhans** (Figure 6.6) These contain two types of cell. Larger α (alpha) cells secrete glucagon, and smaller β (beta) cells secrete insulin.

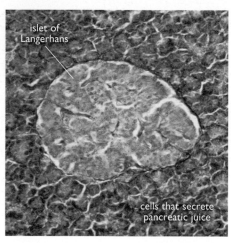

Figure 6.6 *Pancreas tissue including an Islet of Langerhans.*

Another symptom of this kind of diabetes is a constant thirst. This is because the high blood glucose concentration stimulates receptors in the hypothalamus of the brain. These 'thirst' centres are stimulated, so that by drinking, the person will dilute their blood.

Severe diabetes is very serious. If it is untreated, the sufferer loses weight and becomes weak and eventually lapses into a coma and dies.

Glucose in the blood is derived from carbohydrates such as starch in the diet, so mild forms of the disease can be treated by controlling the patient's diet, limiting the amount of carbohydrate that they eat. More serious cases of diabetes need daily injections of insulin to keep the glucose level in the blood at the correct level. People with diabetes can check their blood glucose using a special sensor. They prick their finger and place a drop of blood on a test strip. The strip is then put into the sensor, which gives them an accurate reading of how much glucose is in their blood (Figure 6.8). They can then tell whether or not they need to inject insulin.

Insulin is a protein, and if it were to be taken by mouth in tablet form, it would be broken down in the gut. Instead it is injected into muscle tissue, where it is slowly absorbed into the bloodstream.

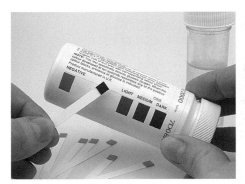

Figure 6.7 *Coloured test strips are used to detect glucose in urine.*

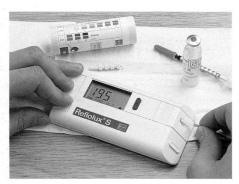

Figure 6.8 *Sensor for measuring blood glucose.*

Insulin for the treatment of diabetes has been available since 1921, and has kept millions of people alive. It was originally extracted from the pancreases of animals such as pigs and cows, and much insulin is still obtained in this way. However, since the 1970s, human insulin has been produced commercially, from genetically modified (GM) bacteria. The bacteria have their DNA 'engineered' to contain the gene for human insulin.

End of Chapter Checklist

You should now be able to:

✓ understand the similarities and differences between the nervous and hormonal systems of communication

✓ understand the action of hormones from:

  • the pituitary (ADH and hormones controlling reproduction)

  • the adrenal glands (adrenaline)

  • the thyroid

  • the pancreas (insulin and glucagon)

  • the testes and ovaries

✓ understand the concept of negative feedback

(At this stage some of these functions are described in outline only. They will be dealt with in more detail in later chapters.)

Questions

1 **a)** *Hormones* are *secreted* by *endocrine glands*. Explain the meaning of the four words in italics.

  b) Identify the hormones A to E in the table.

Hormone	One function of this hormone
A	stimulates the liver to convert glucose to glycogen
B	controls the 'fight or flight' responses
C	in boys, controls the breaking of the voice at puberty
D	completes the development of the uterus lining during the menstrual cycle
E	stimulates an increase in the basal metabolic rate

2 The graph shows the changes in blood glucose in a healthy woman over a 12-hour period.

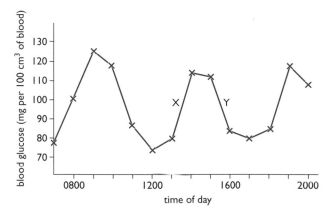

a) Explain why there was a rise in blood glucose at X.

b) How does the body bring about a decrease in blood glucose at Y? Your answer should include the words insulin, liver and pancreas.

c) Diabetes is a disease where the body cannot control the concentration of glucose in the blood.

  i) Why is this dangerous?

  ii) Describe two ways a person with diabetes can monitor their blood glucose level.

  iii) Explain two ways that a person with diabetes can help to control their blood glucose level.

3 **a)** Explain the meaning of the term 'negative feedback'.

  b) Describe the negative feedback loop involving glucagon that takes place in response to a fall in blood glucose.

4 Explain these observations:

  a) Mice that had their thyroid gland removed were less able to survive in freezing conditions than mice with intact thyroid glands.

  b) People living in areas of the world where there is no iodine in the drinking water can develop a condition called a goitre.

Chapter 7: Form and Movement

Humans are **vertebrates** – the human body is supported by an internal skeleton made of bone, with a central 'axis' comprising the **vertebral column**, or backbone, which is where the word 'vertebrate' comes from. Attached to the bones of the skeleton are **voluntary muscles**. These muscles are stimulated to contract by the nervous system, under conscious control by the brain. Contraction of the muscles pulls on bones, producing movement.

The skeleton

The human skeleton (Figure 7.1) has several functions. It protects many vital organs. For example, the **cranium** (skull) protects the brain, eyes and ears, the vertebrae protect the spinal cord, and the ribcage protects the heart and lungs. Bone also makes some components of the blood. Red blood cells and platelets are made in the marrow or larger bones, such as the sternum, femur and pelvis (Figure 7.2). However, the skeleton's most obvious roles are in support and movement.

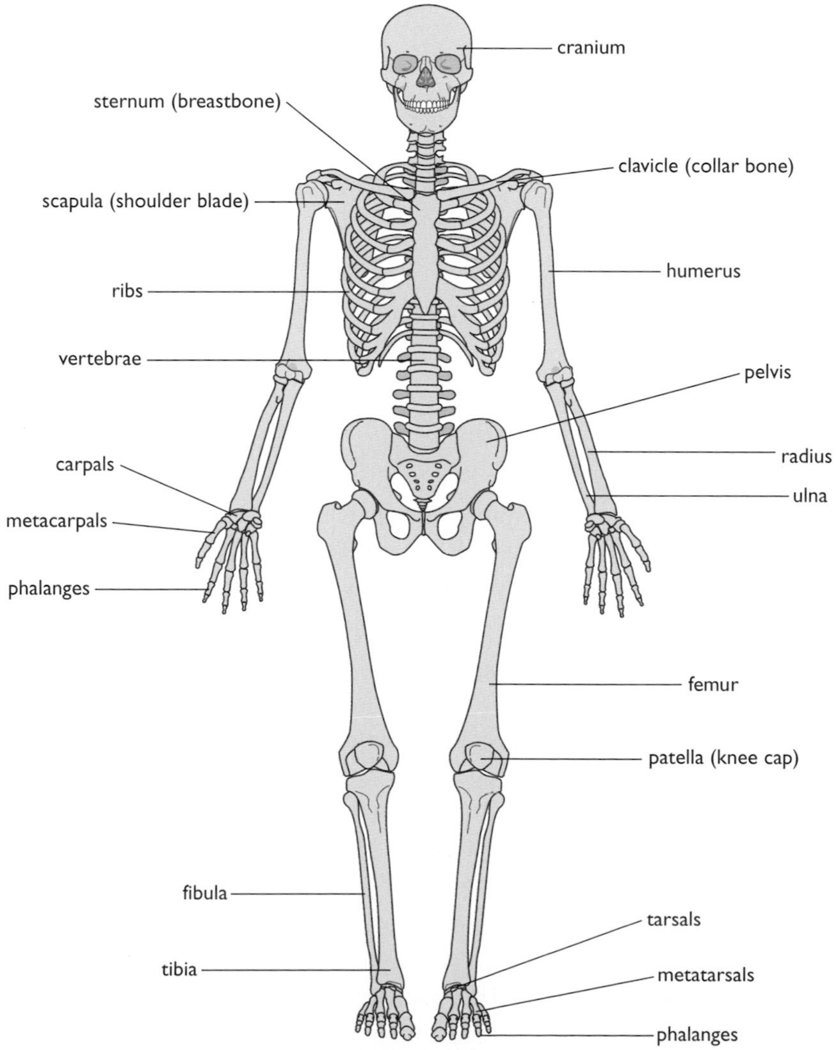

Large animals such as humans need a system of support for their body – a skeleton. The human skeleton not only supports the body and allows movement; it also protects internal organs and makes some components of the blood. This chapter looks at the functions of the skeleton.

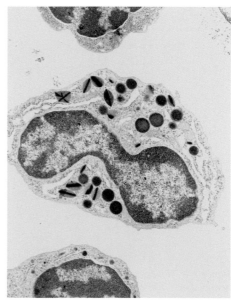

Figure 7.2 *Bone marrow contains cells that divide to produce all our red blood cells and platelets.*

Figure 7.1 *The human skeleton.*

Labels on the skeleton: cranium, sternum (breastbone), clavicle (collar bone), scapula (shoulder blade), humerus, ribs, vertebrae, pelvis, carpals, radius, ulna, metacarpals, phalanges, femur, patella (knee cap), fibula, tarsals, tibia, metatarsals, phalanges

The vertebral column, cranium and ribcage make up the **axial skeleton**. The other bones are attached to this axis – the scapulas, clavicles, pelvis and limbs bones, forming the **appendicular skeleton**.

Bone

Bone is a hard substance because it contains calcium salts, mainly calcium phosphate. This results in a rigid material that can resist bending and compression (squashing) forces. Although out of the body a bone looks dead, in the body it is a living organ, made of cells called **osteocytes**. These cells, along with protein fibres, stop the bone being too brittle. The presence of living cells and blood vessels in the bone also means that bones can repair themselves if they are broken. There are many different shapes and sizes of bones in the human body, from the tiny ear ossicles to long bones like the femur. Figure 7.3 shows a section through a long bone, and Figure 7.4 is a photomicrograph of a section through the outer part of a bone.

If a long bone such as the femur from a chicken is left in a beaker of acid, the acid will dissolve the calcium salts from the bone. The fibres that are left are tough but flexible, so afterwards the bone will bend, and can even be tied in a knot!

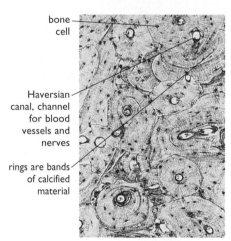

Figure 7.4 *This is a cross-section through a normal bone. The rings are calcified material.*

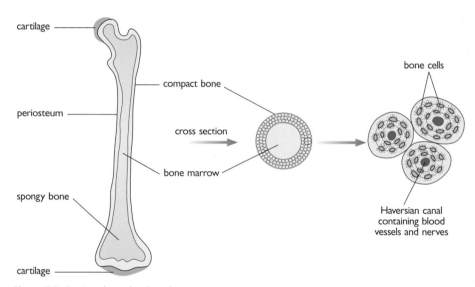

Figure 7.3 *Section through a long bone.*

The middle of a bone is composed of **spongy bone**, with fewer calcium salts, and containing spaces like a sponge, making it less hard. The spaces are filled with bone marrow. Some larger bones have a hollow central cavity containing bone marrow, which makes them lighter. Marrow stores fats and produces blood cells.

The arrangement of the bone matrix in concentric rings makes a long bone very good at resisting compression forces due to the weight of the body, while still allowing the bone a certain amount of side-to-side flexibility.

The outside of the bone is made of harder material, called **compact bone**. In the embryo, bones start off made of cartilage, but as the embryo grows, the cartilage is gradually replaced by bone, a process called **ossification**. This process is nearly complete by the time a baby is born. Osteocytes arrange themselves in rings called **Haversian systems**, around canals containing blood vessels and nerves. The osteocytes secrete calcium phosphate salts, which, along with protein fibres, make up the bone **matrix**.

Cartilage remains present at the ends of long bones, where it acts as a cushion between two bones at a joint (see below). Cartilage is a tough but flexible tissue containing cells called **chondrocytes**. They secrete a matrix containing various types of protein fibres.

Covering the outer surface of the bone is a tough membrane called the **periosteum**.

A person's long bones stop growing in their late teens. During childhood, a healthy balanced diet is essential for bone growth. In particular, calcium is needed for the calcium salts in the bone matrix, and vitamin D is needed to allow calcium to be absorbed by the body. Lack of vitamin D causes the bone disease called rickets (see Chapter 3, page 37). Although bones stop growing in length by adulthood, they continue changing in internal structure throughout adult life, under the influence of hormones, diet, exercise and general health.

Joints

When humans move, the bones move relative to each other. The point where two bones meet is called a **joint**. We say that the bones **articulate** at joints. A movable joint such as the hip or elbow needs to have certain features. These include:

- a way to keep the ends of the bones held together, so that they don't separate (**dislocate**)

- a means of reducing friction between the ends of the moving bones

- a shock-absorbing surface between the two bones.

You can see the structures that do these things in Figure 7.6.

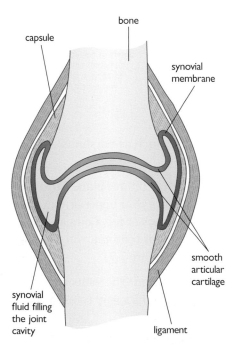

Figure 7.6 *The structure of a synovial joint. The size of the space filled with synovial fluid has been exaggerated in the diagram.*

Bones do not grow in length at their ends. In a child, a long bone has regions of growth just short of the ends of the bone. These regions are made of cartilage. The bone gets longer by the regions of cartilage growing, then becoming ossified.

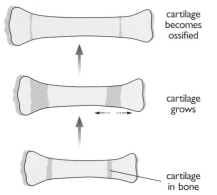

Figure 7.5 *How a long bone grows.*

Osteoporosis is a medical condition that affects elderly people. Their bones lose calcium salts, and become weak and easily broken. Loss of bone like this is normal in all people from middle age, so that by the time a person is aged 70 they will probably have lost about a third of their bone mass. In some individuals it can be particularly bad, especially in women after the menopause, due to changes in hormone levels. There is no cure for osteoporosis, but it can be helped by good diet, calcium and vitamin D supplements, and treatment with hormones.

Movable joints are called **synovial** joints. They contain a liquid called **synovial fluid**, which is secreted by the **synovial membrane**, lining the space in the middle of the joint. Synovial fluid is oily and acts as a lubricant, reducing the friction between the ends of the bones. The end of each bone has an articulating surface covered with a smooth layer of **cartilage**. Cartilage is a strong material, but it is not brittle. It acts as a shock absorber between the ends of the bones, rather like a rubber gym mat compresses to absorb the shock when you fall over.

The correct definition of an 'elastic' material is one which, when you bend or stretch it, will return to its original shape. In this chapter we use the word 'elastic' to mean 'easily stretched'.

The joint is surrounded by a tough fibrous **capsule**, and held together by **ligaments**, which run from one bone to the other across the joint. Ligaments are composed of fibres that make them very tough. They have great strength to resist stretching, called **tensile** strength. However, ligaments have some elasticity, so that they allow joints to bend without the bones becoming dislocated.

There are various kinds of joints in the body. They are classified into three groups by how much movement they allow:

- freely movable joints
- partially movable joints
- immovable (fixed) joints.

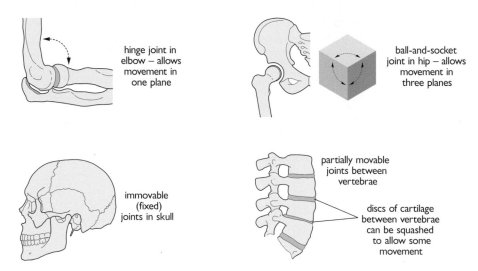

Figure 7.7 *Different types of joint.*

Freely movable joints include **ball and socket** joints, such as the shoulder and hip, and **hinge** joints, such as the elbow.

A ball and socket joint allows movement any direction, in all three planes. Imagine your shoulder joint is at the corner of a box. You can move your upper arm in any of the three planes formed by the sides of the box: side, back or top. Compare this with a hinge joint, such as your elbow – due to the shape of the joint, you can only move this in one plane, as when bending your arm (see Figure 7.7).

Partially movable joints allow a slight degree of movement. The joints between vertebrae are like this. They allow a little movement when the spinal column bends (Figure 7.7).

Some joints are **immovable** or **fixed** joints, for example those between the bones of the skull. The cranium consists of 22 bones, most of them fused together, allowing no movement.

Muscles

The word we use to mean 'not very elastic' is **inelastic**. Both ligaments and tendons have a high tensile strength, but ligaments are fairly elastic, while tendons are inelastic. Ligaments join bone to bone across a joint, while tendons join muscle to bone.

Muscles are organs that are attached to bones and move them by contracting, pulling on the bone. At the end of a muscle there are **tendons**. A tendon attaches the muscle to the bone. Tendons have very high tensile strength, like ligaments, but unlike ligaments they are not very elastic. This means that they don't stretch when the muscle contracts.

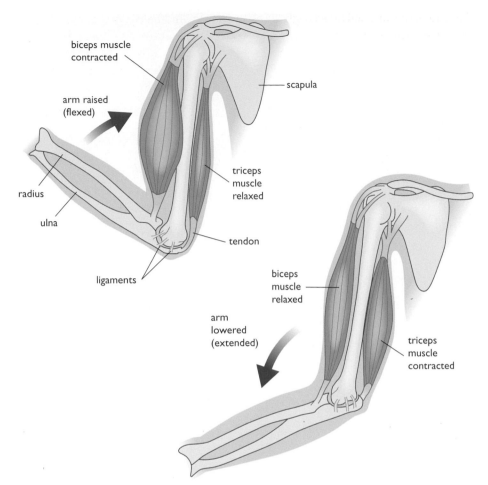

biceps muscle
contracted

scapula

arm raised
(flexed)

triceps
muscle
relaxed

radius

ulna

tendon

ligaments

biceps
muscle
relaxed

arm
lowered
(extended)

triceps
muscle
contracted

Figure 7.8 *The biceps and triceps muscles contract to move the arm at the elbow joint. The biceps flexes the arm, while its antagonistic partner, the triceps, extends the arm.*

Figure 7.9 *A body-builder performing a biceps curl – you can see clearly the contraction of the biceps.*

Muscles cannot push, they can only pull – in other words they are not able to expand actively. When muscles get longer, it is because they are stretched by the contraction of another muscle. When a muscle is being stretched it is relaxed (the opposite of contracted). Because of this, muscles usually work in pairs; one contracting while the other relaxes. These are called **antagonistic pairs**. One of the simplest examples of an antagonistic pair of muscles is the arrangement of the **biceps** and **triceps** muscles in the arm (Figure 7.8).

When the biceps muscle contracts it bends, or **flexes** the arm at the elbow joint. Contraction of the triceps straightens, or **extends** the arm. Of course, there are other muscles in the arm which produce movement in other directions.

When a muscle contracts, the bone at one end of the muscle moves and the bone at the other end stays still. The place where the muscle is attached to the stationary bone is called the **origin**. The place where it is attached to the moving bone is called the **insertion**. When a muscle contracts, the insertion moves towards the origin.

You can identify other antagonistic pairs of muscles in the body. For example, when we run we use several sets of muscles that cause bending at the hip, knee and ankle joints (Figure 7.10).

There are three types of muscle in the body – **skeletal**, **cardiac** and **smooth** muscle. The first of these is the muscle attached to the skeleton. It is described as **voluntary** muscle, because it is under the conscious control of the brain. Cardiac muscle is only found in the heart. It is **involuntary**, meaning not under conscious control. Smooth muscle is also involuntary. It is found in the wall of the gut, bladder, uterus, sperm ducts, blood vessels and other organs.

gluteus (buttocks muscle)

psoas
(groin muscle)

quadriceps (thigh muscle)

biceps femoris

gastrocnemius
(calf muscle)

anterior
tibialis
(shin
muscle)

Figure 7.10 *Antagonistic muscles are used in running. The main muscle that flexes the knee joint is the biceps femoris, while its antagonistic partner, extending the knee, is the quadriceps. Can you work out which muscles flex and extend at the hip and ankle?*

Bending the spine

The movements at the elbow, shoulder, knee and hip are easy to see and understand. Other movements are less obvious, but just as necessary. For example, we need to be able to bend our spine from side to side and from front to back. When this happens, there is not just one point at which the movement takes place, but several.

The spine is sometimes called the **vertebral column** because it is made of vertebrae.

Our spine is made from bones called **vertebrae**. These bones are joined by synovial joints, but the joints cannot bend like the elbow joint. The bones at each joint can be pulled very slightly closer together or further apart to allow a slight bending movement. Because of this slight bending at each joint, the whole spine can bend.

The main muscles responsible for bending the spine from side to side are the **rectus** muscles (Figure 7.11).

The vertebrae in different regions of the spine are different. Their shape and size is related to the kind of load they have to bear (Figure 7.12).

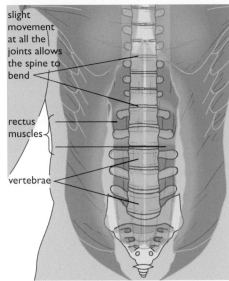

Figure 7.11 *Contraction and relaxation of the rectus muscles produces limited movement at each joint in the spine.*

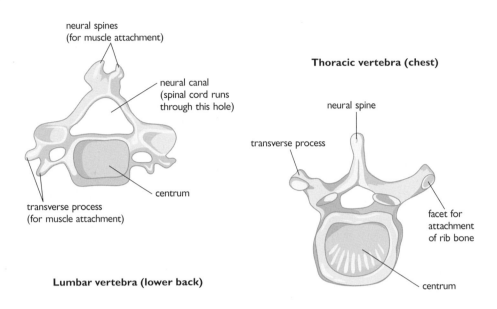

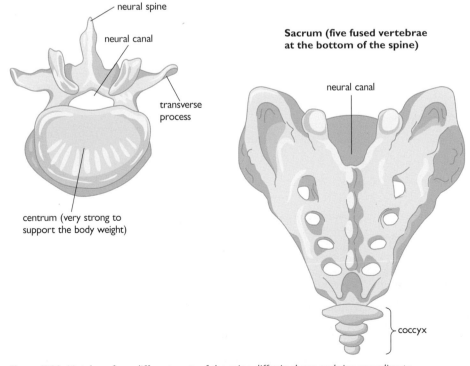

Figure 7.12 *Vertebrae from different parts of the spine differ in shape and size according to their position.*

Muscle contraction and exercise

Skeletal muscle is made up of highly specialised muscle cells or **fibres**, arranged in bundles in a connective tissue sheath. Muscle fibres are adapted for contraction. Under very high magnification, using an electron microscope, we can see that they are composed of fine protein filaments (Figure 7.13).

There are two types of filament, thick and thin. When a muscle contracts, the thin filaments slide past the thick filaments, making the fibres shorter.

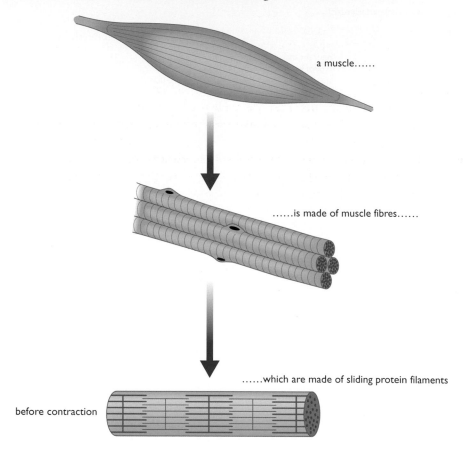

a muscle......

......is made of muscle fibres......

......which are made of sliding protein filaments

before contraction

after contraction

Figure 7.13 *Muscles are composed of muscle fibres, made of fine protein filaments. When a muscle contracts, the filaments slide over each other.*

Contraction of muscle fibres needs energy; this comes from respiration (Chapter 1). Blood vessels supply glucose and oxygen to the muscle fibres. The fibres respire, converting the glucose and oxygen into carbon dioxide and water. Energy is released for the fibres to contract, but the process is not 100% efficient, and some energy is lost as heat. When we carry out strenuous exercise, our muscles demand a greater supply of glucose and oxygen than usual. In addition, carbon dioxide and heat need to be removed at a faster rate. To achieve this, various changes take place in the body.

A healthy balanced diet is not just needed for bone growth. Muscles are largely made of protein, and a balanced diet containing an adequate supply of amino acids is needed for protein synthesis in muscles.

- The breathing rate increases, so more oxygen is taken into the blood by the lungs, and more carbon dioxide is lost. The volume of each breath also increases.

- The heart rate increases, pumping more oxygenated blood to the muscles.

- Blood is diverted away from places like the gut, and towards the muscles.

- The skin carries out processes such as vasodilation and sweating (see Chapter 8) which remove excess heat from the body.

Even when a muscle is relaxed, some of its fibres are contracted. This state of partial contraction is called muscle **tone**. It keeps our muscles taut, but not enough to cause movement. Muscle tone keeps us upright when we are standing or sitting. Not all the muscle fibres are contracted at once – they take it in turn, in relays.

Regular exercise keeps muscles toned. This has other benefits. The partial tension in the muscle fibres means that they are ready to contract more quickly. Exercise also develops muscles, because they make more protein filaments, and are able to produce a stronger contraction. If you never exercise, and then have a sudden 'work out', your muscles will feel sore and stiff afterwards. You can avoid this by regular exercise, which also improves the circulation to the muscles, heart and lungs, and keeps the joints working smoothly.

End of Chapter Checklist

You should now be able to:

✓ recall the main parts of the axial and appendicular skeleton, including the vertebral column, ribcage, skull, scapula, clavicle, pelvis and limb bones

✓ describe the functions of the skeleton

✓ recall the structure of a long bone

✓ describe the structure of a synovial joint

✓ explain the functions of joints, using the elbow, shoulder and cartilaginous intervertebral joint as examples

✓ explain the relationship between voluntary muscles and bones to bring about movement, illustrated by the biceps and triceps muscles and associated bones in the arm and shoulder

✓ recall the dietary factors controlling the healthy development of muscle and bone

✓ understand the effects of exercise and the benefits of regular exercise on muscles

Questions

1 **a)** Which parts of the body are protected by

    i)   the cranium

    ii) the vertebral column

    iii) the ribs?

  b) Between the vertebrae are discs of cartilage. From your knowledge of the properties of cartilage, suggest what their function is.

  c) Name three components of bone.

2 The synovial membrane, synovial fluid and ligamants are parts of a synovial joint. Explain the function of each.

3 Copy and complete the following paragraph, putting the most suitable word or words in the spaces:

Muscles are normally found in .... pairs, such as the triceps and biceps of the arm. When the .... contracts, it flexes the arm, whereas when the .... contracts, it straightens or extends the arm. When muscles are exercised, they need an increased supply of energy. This is supplied by the process of cell .... , which uses .... and .... and makes carbon dioxide. Exercise also produces heat.

4 Write an essay about the advantages of taking regular exercise. This chapter has outlined some of the benefits, but you can research this important topic more thoroughly and summarise the effects on the muscles, joints, heart and blood system, and the lungs. The length of your essay should be about two sides of writing, plus diagrams. Use other books or the Internet as sources of information.

5 The diagram shows the muscles and bones in a human forearm when the forearm is being raised.

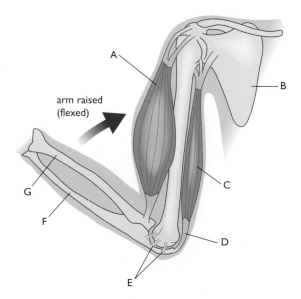

arm raised (flexed)

a) Name the parts labelled A, B, C, D, E, F and G.

b) A and C are antagonistic muscles. What does this mean?

c) i) Describe the function of the structures labelled E.

   ii) What properties of structures E adapt them to this function?

d) What must happen for the forearm to be lowered? Explain your answer.

6 If a muscle is isolated and stimulated by giving it a brief electrical shock, the muscle responds by 'twitching'. This is shown in the graph.

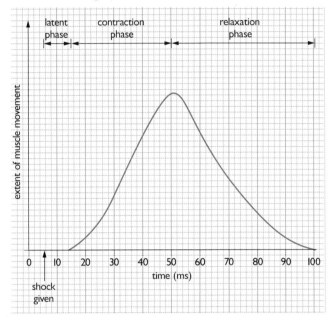

a) In this experiment, how long did each of the three phases of the twitch response last?

b) For how long could the muscle have been exerting a force? Explain your answer.

c) Muscles often exist in antagonistic pairs. Explain how antagonistic muscles can:

   i) bend and straighten the arm

   ii) keep the body upright and straight.

d) Muscles are made from muscle fibres. There are three types of muscle fibre: type I, type IIa and type IIb. The table gives some of the properties of each type.

Property	Fibre type		
	I	**IIa**	**IIb**
resistance to fatigue	high	moderate	low
aerobic/ anaerobic respiration	aerobic	both	anaerobic
speed of contraction	slow	intermediate	fast
tension produced when contracted	low	intermediate	high

Different athletes have different combinations of the three types of fibres in their muscles. Describe the combination of fibres you would expect to find in:

   i) an endurance athlete (such as a marathon runner)

   ii) a power athlete (such as a weightlifter).

e) Explain your answers to d) parts i) and ii).

Chapter 8: Homeostasis and Excretion

The kidneys play a major part in homeostasis (maintaining a balance of substances in the body) and excretion (removal of waste products from cell metabolism). This chapter is mainly concerned with the activities of the kidneys. It also deals with another important aspect of homeostasis, that of maintaining a steady body temperature.

Inside our bodies, conditions are kept relatively constant. This is called **homeostasis**. The kidneys are organs which have a major role to play in both homeostasis and in the removal of waste products, or **excretion**. They filter the blood, removing substances and controlling the concentration of water and solutes in the blood and other body fluids.

Homeostasis

If you were to drink a litre of water and wait for half an hour, your body would soon respond to this change by producing about the same volume of urine. In other words, it would automatically balance your water input and water loss. Drinking is the main way that our bodies gain water, but there are other sources (Figure 8.1). Some water is present in the food that we eat, and a small amount is formed by cell respiration. The body also loses water, mostly in urine, but also smaller volumes in sweat, faeces and exhaled air. Every day, we gain and lose about the same volume of water, so that the total content of our bodies stays more or less the same. This is an example of homeostasis. The word 'homeostasis' means 'steady state', and refers to keeping conditions inside the body relatively constant.

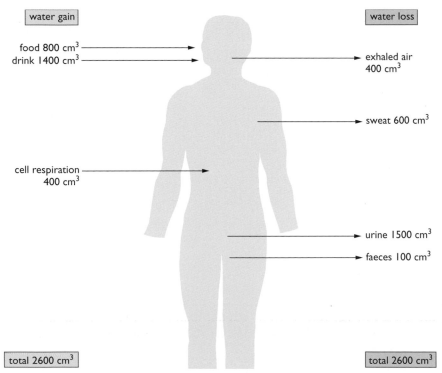

water gain		water loss
food 800 cm³		exhaled air 400 cm³
drink 1400 cm³		sweat 600 cm³
cell respiration 400 cm³		urine 1500 cm³
		faeces 100 cm³
total 2600 cm³		total 2600 cm³

Figure 8.1 *The daily water balance of an adult.*

Inside the body is known as the **internal environment**. The environment means the surroundings of an organism. The internal environment is the surroundings of the cells inside the body. Organisms can respond to changes in their internal environment, just as they respond to changes in their surroundings.

Chapter 8: Homeostasis and Excretion

The internal environment refers to the blood, together with the tissue fluid (see Chapter 4, page 60).

Tissue fluid is a watery solution of salts, glucose and other solutes. It surrounds all the cells of the body, forming a pathway for the transfer of nutrients between the blood and the cells. Tissue fluid is formed by leakage from blood capillaries. It is similar in composition to blood plasma, but lacks the plasma proteins.

It is not just water and salts that are kept constant in the body. Many other components of the internal environment are maintained. For example, the level of carbon dioxide in the blood is regulated, along with the blood pH, the concentration of dissolved glucose (see Chapter 6) and the body temperature.

Homeostasis is important because cells will only function properly if they are bathed in a tissue fluid which provides them with their optimum conditions. For instance, if the tissue fluid contains too many solutes, the cells will lose water by osmosis, and become dehydrated. If the tissue fluid is too dilute, the cells will swell up with water. Both conditions will prevent them working efficiently and might cause permanent damage. If the pH of the tissue fluid is not correct, it will affect the activity of the cell's enzymes, as will a body temperature much different from 37 °C. It is also important that excretory products are removed. Substances such as urea must be prevented from building up in the blood and tissue fluid, where they would be toxic to cells.

> Homeostasis means 'keeping the conditions in the internal environment of the body relatively constant'. Keeping the water and salt content of the internal environment constant is known as **osmoregulation**.

Urine

An adult human produces about 1.5 dm³ of urine every day, although this volume depends very much on the amount of water drunk and the volume lost in other forms, such as sweat. Every litre of urine contains about 40 g of waste products and salts (Table 8.1).

> 'Salts' in urine or in the blood are present as ions. For example, the sodium chloride in Table 8.1 will be in solution as sodium ions (Na^+) and chloride ions (Cl^-). Urine contains many other ions, such as potassium (K^+), phosphate (HPO_4^{2-}) and ammonium (NH_4^+), and removes excess ions from the blood.

Substance	Amount (g/dm³)
urea	23.3
ammonia	0.4
other nitrogenous waste	1.6
sodium chloride (salt)	10.0
potassium	1.3
phosphate	2.3

Table 8.1: *Some of the main dissolved substances in urine.*

Notice the words **nitrogenous waste**. Urea and ammonia are two examples of nitrogenous waste. It means that they contain the element **nitrogen**. All animals have to excrete a nitrogenous waste product.

The reason behind this is quite involved. Carbohydrates and fats only contain the elements carbon, hydrogen and oxygen. Proteins, on the other hand, also contain nitrogen. If the body has too much carbohydrate or fat, these substances can be stored, for example as glycogen in the liver, or as fat under the skin and around other organs. Excess proteins, or their building blocks (called amino acids) *cannot* be stored. The amino acids are first broken down in the liver. They are converted into

Excretion is the process by which waste products of metabolism are removed from the body. In humans, the main nitrogenous excretory substance is urea. Another excretory product is carbon dioxide, produced by cell respiration and excreted by the lungs (Chapter 2). The human skin is also an excretory organ, since the sweat that it secretes contains small amounts of urea.

carbohydrate (which is stored as glycogen) and the main nitrogen-containing waste product, urea. The urea passes into the blood, to be filtered out by the kidneys during the formation of urine. Notice that the urea is made by chemical reactions in the cells of the body (the body's metabolism). 'Excretion' means getting rid of waste of this kind. When the body gets rid of solid waste from the digestive system (faeces), this is not excretion, since it contains few products of *metabolism*, just the 'left over remains' of undigested food, along with bacteria and dead cells.

So the kidney is really carrying out two functions. It is a *homeostatic* organ, controlling the water and salt (ion) concentration in the body as well as an *excretory* organ, concentrating nitrogenous waste in a form that can be eliminated.

The urinary system

The human urinary system is shown in Figure 8.2.

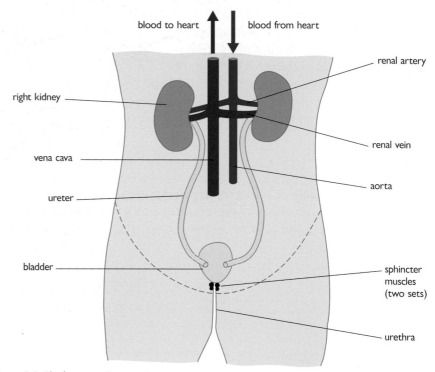

Figure 8.2 *The human urinary system.*

Each kidney is supplied with blood through a short **renal artery**. This leads straight from the body's main artery, the aorta, so the blood entering the kidney is at a high pressure. Inside each kidney the blood is filtered, and the 'cleaned' blood passes out through each **renal vein** to the main vein, or vena cava. The urine passes out of the kidneys through two tubes, the **ureters**, and is stored in a muscular bag called the **bladder**.

A baby cannot control its voluntary sphincter. When the bladder is full, the baby's involuntary sphincter relaxes, releasing the urine. A toddler learns to control this muscle and hold back the urine.

The bladder has a tube leading to the outside, called the **urethra**. The wall of the urethra contains two ring-like muscles, called **sphincters**. They can contract to close the urethra and hold back the urine. The lower sphincter muscle is consciously controlled, or voluntary, while the upper one is involuntary – it automatically relaxes when the bladder is full.

The kidneys

If you cut a kidney lengthwise as in Figure 8.3 you should be able to find the structures shown.

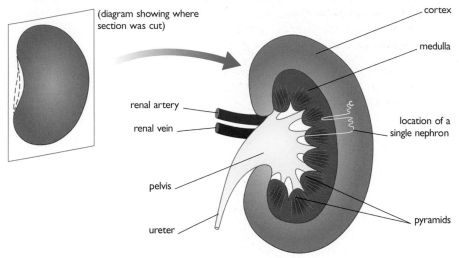

Figure 8.3 *Section through a kidney cut along the plane shown.*

There is not much that you can make out without the help of a microscope. The darker outer region is called the **cortex**. This contains many tiny blood vessels that branch from the renal artery. It also contains microscopic tubes that are not blood vessels. They are the filtering units, called **kidney tubules** or **nephrons** (from the Greek word *nephros*, meaning kidney). The tubules then run down through the middle layer of the kidney, called the **medulla**. The medulla has bulges called **pyramids** pointing inwards towards the concave side of the kidney. The tubules in the medulla eventually join up and lead to the tips of these pyramids, where they empty urine into a space called the **pelvis**. The pelvis connects with the **ureter**, carrying the urine to the **bladder**.

By careful dissection, biologists have been able to find out the structure of a single tubule and its blood supply (Figure 8.4). There are about a million of these in each kidney.

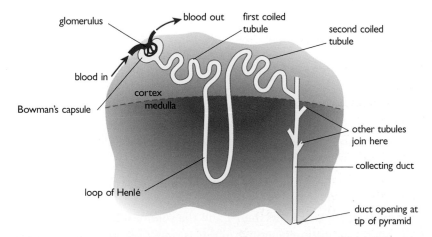

Figure 8.4 *A single nephron, showing its position in the kidney. Each kidney contains about a million of these filtering units.*

In some biology books the first and second coiled tubules are called the **proximal** and **distal convoluted** tubules. 'Proximal' means near the start of the tubule, and 'distal' means near the end.

At the start of the nephron is a hollow cup of cells called the **Bowman's capsule**. It surrounds a ball of blood capillaries called a **glomerulus** (the plural is glomeruli). It is here that the blood is filtered. Blood enters the kidney through the renal artery, which divides into smaller and smaller arteries. The smallest arteries, called **arterioles**, supply the capillaries of the glomerulus (Figure 8.5).

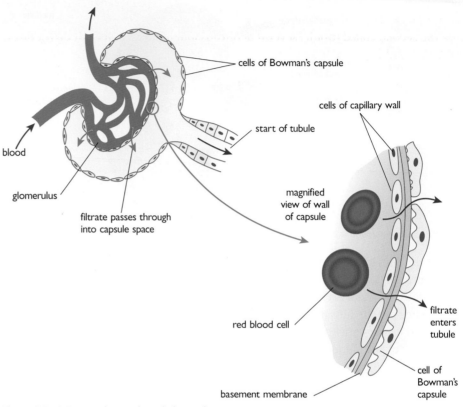

Figure 8.5 *A Bowman's capsule and glomerulus.*

The cells of the glomerulus capillaries do not fit together very tightly, there are spaces between them making the capillary walls much more permeable than others in the body. The cells of the Bowman's capsule also have gaps between them, so only act as a coarse filter. It is the basement membrane which is the fine molecular filter.

A blood vessel with a smaller diameter carries blood away from the glomerulus, leading to capillary networks which surround the other parts of the nephron. Because of the resistance to flow caused by the glomerulus, the pressure of the blood in the arteriole leading to the glomerulus is very high. This pressure forces fluid from the blood through the walls of the capillaries and the Bowman's capsule, into the space in the middle of the capsule. Blood in the glomerulus and the space in the capsule are separated by two layers of cells, the capillary wall and the wall of the capsule. Between the two cell layers is a third layer called the **basement membrane**, which is not made of cells. These layers act like a filter, allowing water, ions and small molecules to pass through, but holding back blood cells and large molecules such as proteins. The fluid that enters the capsule space is called the **glomerular filtrate**. This process, where the filter separates different sized molecules under pressure, is called **ultrafiltration**.

The kidneys produce about $125\,cm^3$ ($0.125\,dm^3$) of glomerular filtrate per minute. This works out at $180\,dm^3$ per day. Remember though, only $1.5\,dm^3$ of urine is lost from the body every day, which is less than 1% of the volume filtered through the capsules. The other 99% of the glomerular filtrate is *reabsorbed* back into the blood.

We know this because scientists have actually analysed samples of fluid from the space in the middle of the nephron. Despite the diameter of the space being only 20 μm (0.02 mm), it is possible to pierce the tubule with microscopic glass pipettes and extract the fluid for analysis. Figure 8.6 shows the structure of the nephron and the surrounding blood vessels in more detail.

There are two **coiled regions** of the tubule in the cortex, separated by a U-shaped loop that runs down into the medulla of the kidney, called the **loop of Henlé**. After the second coiled tubule, several nephrons join up to form a **collecting duct**, where the final urine passes out into the pelvis.

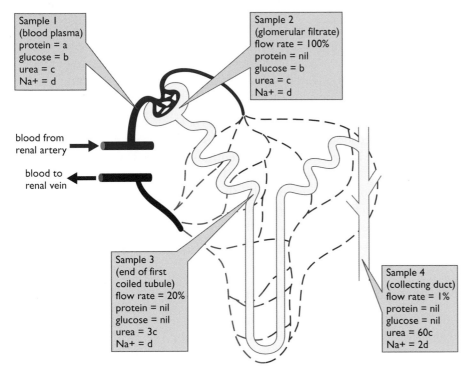

Figure 8.6 *A nephron and its blood supply. Samples 1–4 show what is happening to the fluid as it travels along the nephron.*

Samples 1–4 show the results of analysing the blood before it enters the glomerulus, and the fluid at three points inside the tubule. The flow rate is a measure of how much water is in the tubule. If the flow rate falls from 100% to 50%, this is because 50% of the water in the tubule has gone back into the blood. To make the explanation easier, the concentrations of dissolved protein, glucose, urea and sodium are shown by different letters (a to d). You can tell the relative concentration of one substance at different points along the tubule from this. For example, urea at a concentration '3c' is three times more concentrated than when it is 'c'.

In the blood (sample 1) the plasma contains many dissolved solutes, including protein, glucose, urea and salts (just sodium ions, Na⁺, are shown here). As we saw above, protein molecules are too big to pass through into the tubule, so the protein concentration in sample 2 is zero. The other substances are at the same concentration as in the blood.

Now look at sample 3, taken at the end of the first coiled part of the tubule. The flow rate that was 100% is now 20%. This must mean that 80% of the water in the tubule has been reabsorbed back into the blood. If no solutes were reabsorbed

The kidney tubule reabsorbs different amounts of various substances. This is called **selective reabsorption**.

Here is a summary of what happens in the kidney nephron.

Part of the plasma leaves the blood in the Bowman's capsule and enters the nephron. The filtrate consists of water and small molecules. As the fluid passes along the nephron, all the glucose is absorbed back into the blood in the first coiled part of the tubule, along with most of the sodium and chloride ions. In the rest of the tubule, more water and ions are reabsorbed, and some solutes like ammonium ions are secreted into the tubule. The final urine contains urea at a much higher concentration than in the blood. It also contains controlled quantities of water and ions.

As well as causing the pituitary gland to release ADH, the receptor cells in the hypothalamus also stimulate a 'thirst centre' in the brain. This makes the person feel thirsty, so that they will drink water, diluting the blood.

along with the water, their concentrations should be *five times* what they were in sample 2. Since the concentration of sodium hasn't changed, 80% of this substance must have been reabsorbed (and some of the urea too). However, the glucose concentration is now zero – *all* of the glucose is taken back into the blood in the first coiled tubule. This is necessary because glucose is a useful substance that is needed by the body.

Finally, look at sample 4. By the time the fluid passes through the collecting duct, its flow rate is only 1%. This is because 99% of the water has been reabsorbed. Protein and glucose are still zero, but most of the urea is still in the fluid. The level of sodium is only 2d, so not all of it has been reabsorbed, but it is still twice as concentrated as in the blood.

This description has only looked at a few of the more important substances. Other solutes are concentrated in the urine by different amounts. Some, like ammonium ions, are secreted *into* the fluid as it passes along the tubule. The concentration of ammonium ions in the urine is about 150 times what it is in the blood.

You might be wondering what the role of the loop of Henlé is. The full answer to this is too complicated for an IGCSE textbook, so a simple explanation will have to be sufficient for now. It is involved with concentrating the fluid in the tubule by causing more water to be reabsorbed into the blood. Mammals with long loops of Henlé can make a more concentrated urine than ones with short loops. Desert animals have many long loops of Henlé, so they are able to produce very concentrated urine, conserving water in their bodies. Animals which have easy access to water, such as otters or beavers, have short loops of Henlé. Humans have a mixture of long and short loops.

Control of the body's water content

Not only can the kidney produce urine that is more concentrated than the blood, it can also *control* the concentration of the urine, and so *regulate* the water content of the blood. This chapter began by asking you to think what would happen if you drank a litre of water. The kidneys respond to this 'upset' to the body's water balance by making a larger volume of more dilute urine. Conversely, if the blood becomes too concentrated, the kidneys produce a smaller volume of urine. These changes are controlled by a hormone produced by the pituitary gland, at the base of the brain. The hormone is called **antidiuretic hormone**, or **ADH**.

'Diuresis' means the flow of urine from the body, so 'antidiuresis' means producing less urine. ADH starts to work when your body loses too much water, for example if you are sweating heavily and not replacing lost water by drinking.

The loss of water means that the concentration of the blood starts to increase. This is detected by special cells in a region of the brain called the **hypothalamus** (see Chapter 5). These cells are sensitive to the solute concentration of the blood, and cause the pituitary gland to release more ADH. The ADH travels in the bloodstream to the kidney. At the kidney tubules, it causes the collecting ducts to become more permeable to water, so that more water is reabsorbed back into the blood. This makes the urine more concentrated, so that the body loses less water and the blood becomes more dilute.

When the water content of the blood returns to normal, this acts as a signal to 'switch off' the release of ADH. The kidney tubules then reabsorb less water.

Similarly, if someone drinks a large volume of water, the blood will become too dilute. This leads to lower levels of ADH secretion, the kidney tubules become less permeable to water, and more water passes out of the body in the urine. In this way, through the action of ADH, the level of water in the internal environment is kept constant.

Oral rehydration therapy

The correct balance of water and ions in the blood and tissue fluid is essential for the healthy functioning of the body. A simple gut infection by any of a number of harmful microorganisms can result in diarrhoea and vomiting. If these symptoms continue, the patient's tissues may become dehydrated and the normal balance of salts upset. In severe cases this leads to brain and kidney damage, and eventually death.

It is a sad fact that, despite all the advances of modern medicine, one of the biggest killers in the world today is diarrhoea. This particularly affects babies and young children, who are not able to adjust to rapid dehydration. In many less developed countries around the world, diarrhoea still kills millions of people every year.

There is a simple answer that saves lives – the **oral rehydration** method. This consists of a pack containing a solution of salts and glucose, which is mixed with sterile water (Figure 8.8). If a child is suffering from diarrhoea, they are fed the solution (Figure 8.9). This helps to rehydrate their tissues, and can prevent further damage or death, even if the diarrhoea has not stopped.

Oral rehydration therapy was first used in the 1970s. As a result, by 2009 the number of infant deaths around the world from diarrhoea had been cut from 5 million to 1.5 million. Medical workers aim to reduce this figure further still, by providing easier access to this cheap, simply prepared treatment.

The action of ADH is another example of **negative feedback** (see Chapter 6). A change in conditions in the body is detected, and starts a process which works to return conditions to normal. When the conditions are returned to normal, the corrective process is switched off (Figure 8.7).

In the situation described, the blood becomes too concentrated. This switches on ADH release, which acts at the kidneys to correct the problem. The word 'negative' means that the process works to eliminate the change. When the blood returns to normal, ADH release is switched off. The feedback pathway forms a 'closed loop'. Many conditions in the body are regulated by negative feedback loops like this.

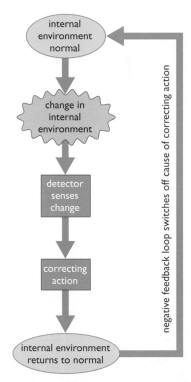

Figure 8.7 In homeostasis, the extent of a correction is monitored by negative feedback.

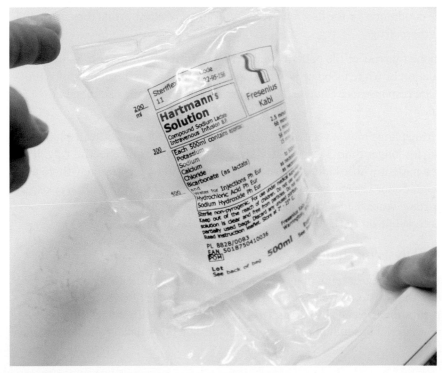

Figure 8.8 An oral rehydration pack.

Figure 8.9 An infant being given oral treatment for dehydration.

Kidney failure – what happens when things go wrong

A kidney can stop working as a result of disease or an accident. We can live perfectly happily with only one kidney, but if both stop working we will die within a week or so, because poisonous waste builds up in the blood. If a person's kidneys fail, there are two ways they can be kept alive. One is to carry out a kidney **transplant**, and the other is for their blood to be filtered through an artificial kidney machine, in a process called **renal dialysis**.

Kidney transplants

A kidney transplant is an operation where a patient receives another person's kidney. This is often donated by a close relative, or it may be from a person who has had a fatal accident (Figure 8.10).

A kidney transplant is a straightforward operation, and kidneys were one of the first human organs to be successfully transplanted. The main problem comes soon after the operation, when the new kidney can be **rejected** by the patient's immune system (see Chapter 13). The immune system recognises the new kidney as being made of 'foreign' tissues and tries to destroy it. There are a number of ways doctors can try to stop the kidney being rejected:

Figure 8.10 *'A kidney being made ready for a transplant operation'.*

- The donated organ is taken from a person with tissues that are as genetically similar to the patient's as possible. This is done by **tissue typing**, which classifies the patient's tissues and tries to find a close 'match' with possible donors' tissues. This is why close relatives are often used as donors – they are more likely to have similar tissues. The ideal donor is an identical twin!

- The patient can be given **immuno-suppressant** drugs, until the kidney is 'accepted' by the body. These are drugs which stop the immune system working properly, so that the kidney is not rejected. However, it leaves the patient open to catching infectious diseases. Transplant patients are kept in hospital under very sterile conditions for the first few weeks after the operation, while they are being given these drugs.

- The bone marrow of the patient can be treated with radiation. This reduces the production of white blood cells, which are the cells that are responsible for the immune response (see Chapter 13). Again, this increases the likelihood of infections.

Figure 8.11 *If a person is killed in an accident and they carry a donor card, it shows the hospital that they have agreed to allow their organs to be used for transplants. It is important that a transplant is carried out as soon as possible after a person has died.*

Kidney transplants have a very high success rate, with over 80% of transplants surviving for longer than 3 years. They are the most simple and effective way of treating a patient whose kidneys have failed permanently. Unfortunately there are not enough donor organs available, despite the number of people who carry organ donor cards (Figure 8.11). Many people are on a waiting list for a kidney transplant. While on the list, they are treated by renal dialysis.

Kidney dialysis machines

The artificial kidney, or renal dialysis machine, filters the patient's blood, removing urea and other waste, as well as excess water and salts. The filter is a special **dialysis membrane** called **Visking tubing**, which looks rather like cellophane. This thin material has millions of tiny holes in it. The holes will let small molecules like water, ions and urea pass through, but not larger molecules such as proteins, or blood cells.

Blood from the patient flows on one side of the membrane, and a watery liquid called **dialysis solution** flows past the other side, in the opposite direction (Figure 8.12). This is a solution of salts and glucose in exactly the concentrations that the body needs.

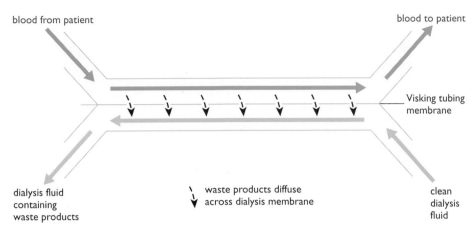

blood from patient

blood to patient

Visking tubing membrane

dialysis fluid containing waste products

waste products diffuse across dialysis membrane

clean dialysis fluid

Figure 8.12 *The principle of renal dialysis. The dialysis membrane filters the blood, removing toxic waste.*

As the blood flows past the membrane, urea and unwanted water and salts diffuse through the holes in the membrane into the dialysis fluid. Cells and large molecules such as proteins are kept back in the blood. The dialysis fluid is replaced with fresh solution all the time, so that after several hours, the patient's blood has been 'cleaned' of toxic waste and the correct balance of water and salts established.

The surface area of the dialysis membrane separating the blood and dialysis fluid must be large to filter the blood enough. To achieve this, the membrane can be arranged in different ways. In some dialysis machines it is in the form of many long narrow tubes, while in others it is arranged as a stack of flat sheets.

In order to carry out dialysis, it is easier to take blood from a vein than an artery, because veins are closer to the skin, and have a wider diameter than arteries (see Chapter 4). However, the blood pressure in veins is too low, so an operation is first carried out to join an artery to a vein, which raises the blood pressure. A tube is then permanently connected to the vein, so that the patient can be linked to the machine without having to use a needle each time. The purified blood returns to the patient through a second tube joined to the vein, and the 'used' dialysis fluid is discarded.

The kidney machine is a complex and expensive piece of apparatus (Figure 8.13). It has pumps to keep the blood and dialysis fluid flowing, traps to prevent air bubbles getting into the blood, and the oxygenation and temperature of the blood is controlled. Although dialysis will keep a person whose kidneys have failed alive, it is a time-consuming and unpleasant process. The patient has to have their blood 'cleaned' for many hours, two or three times a week. A transplant, if one becomes available, is a much better option.

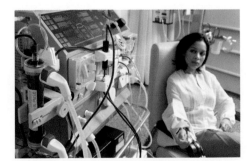

Figure 8.13 *A patient connected to a kidney dialysis machine.*

Control of body temperature

You may have heard mammals and birds described as 'warm blooded'. A better word for this is **homeothermic**. It means that they keep their body temperature constant, despite changes in the temperature of their surroundings. For example, the body temperature of humans is kept steady at about 37 °C, give or take a few tenths of a degree. This is another example of homeostasis. All other animals are 'cold blooded'. For example, if a lizard is kept in an aquarium at 20 °C, its body temperature will be 20 °C too. If the temperature of the aquarium is raised to 25 °C, the lizard's body temperature will rise to 25 °C as well. We can show this difference between homeotherms and other animals as a graph (Figure 8.14).

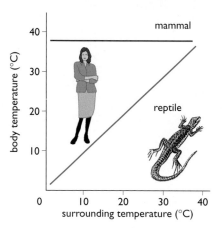

Figure 8.14 *The temperature of a homeotherm, such as a mammal, is kept constant at different external temperatures, whereas the lizard's body temperature changes.*

In the wild, lizards keep their temperature more constant than in Figure 8.14, by adapting their behaviour. For example, in the morning they may bask in the sun to warm their bodies, or at midday, if the sun is too hot, retreat to holes in the ground to cool down.

The real difference between homeotherms and all other animals is that homeotherms can keep their temperatures constant by using **physiological** changes for generating or losing heat. For this reason, mammals and birds are also called **endotherms**, meaning 'heat from inside'.

An endotherm uses heat from the chemical reactions in its cells to warm its body. It then controls its heat loss by regulating processes like sweating and blood flow through the skin. Endotherms use behavioural ways to control their temperature too. For example, penguins 'huddle' to keep warm, and humans put on extra clothes in winter.

What is the advantage of a human maintaining a body temperature of 37 °C? It means that all the chemical reactions taking place in the cells of the body can go on at a steady, predictable rate. The metabolism doesn't slow down in cold environments. If you watch goldfish in a garden pond, you will notice that in summer, when the pond water is warm, they are very active, swimming about quickly. In winter, when the temperature drops, the fish slow down and become very sluggish in their actions. This would happen to mammals too, if their body temperature was not kept steady.

It is also important that the body does not become *too* hot. The cells' enzymes work best at 37 °C. At higher temperatures enzymes, like all proteins, are destroyed by **denaturing** (see Chapter 1).

Physiology is a branch of biology that deals with how the bodies of animals or plants work, for example how muscles contract, how nerves send impulses, or how xylem carries water through plants. In this chapter you have read about kidney physiology.

Monitoring body temperature

In humans and other mammals the core body temperature is monitored by a part of the brain called the **thermoregulatory centre**. This is located in the hypothalamus of the brain (see Figure 5.17, page 82). It acts as the body's thermostat.

If a person goes into a warm or cold environment, the first thing that happens is that temperature receptors in the skin send electrical impulses to the hypothalamus, which stimulates the brain to alter our behaviour. We start to feel hot or cold, and usually do something about it, such as finding shade or having a cold drink.

If changes to our behaviour are not enough to keep our body temperature constant, the thermoregulatory centre in the hypothalamus detects a change in the temperature of the blood flowing through it. It then sends signals via nerves to other organs of the body, which regulate the temperature by physiological means.

Control of body temperature is another example of a negative feedback system. The thermoregulatory centre detects a change in body temperature and brings about mechanisms to correct the change and bring the temperature back to normal.

A **thermostat** is a switch that is turned on or off by a change in temperature. It is used in electrical appliances to keep their temperature steady. For example, a thermostat in an iron can be set to 'hot' or 'cool' to keep the temperature of the iron set for ironing different materials.

The skin and temperature control

The human skin has a number of functions related to the fact that it forms the outer surface of the body. These include:

- forming a tough outer layer able to resist mechanical damage
- acting as a barrier to the entry of disease-causing microorganisms
- forming an impermeable surface, preventing loss of water
- acting as a sense organ for touch and temperature changes
- controlling the loss of heat through the body surface.

Figure 8.15 shows the structure of human skin. It is made up of three layers, the epidermis, dermis and hypodermis.

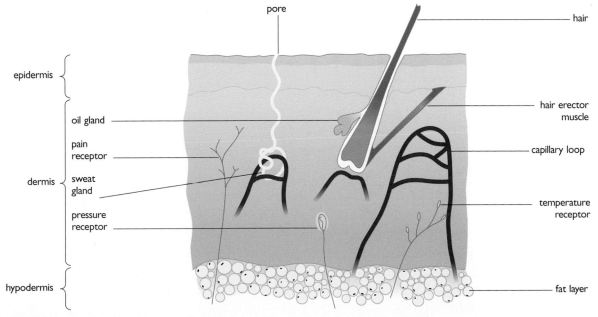

Figure 8.15 *A section through human skin.*

The outer **epidermis** consists of dead cells that stop water loss and protect the body against invasion by microorganisms such as bacteria. The **hypodermis** contains fatty tissue, which insulates the body against heat loss and is a store of energy. The middle layer, the **dermis**, contains many sensory receptors. It is also the location of sweat glands and many small blood vessels, as well as hair follicles. These last three structures are involved in temperature control.

Imagine that the hypothalamus detects a rise in the central (core) body temperature. Immediately it sends nerve impulses to the skin. These bring about changes to correct the rise in temperature.

First of all, the **sweat glands** produce greater amounts of sweat. This liquid is secreted onto the surface of the skin. When a liquid evaporates, it turns into a gas. This change needs energy, called the **latent heat of vaporisation**. When sweat evaporates, the energy is supplied by the body's heat, cooling the body down. It is not that the sweat is cool – it is secreted at body temperature. It only has a cooling action when it evaporates. In very humid atmospheres (e.g. a tropical rainforest) the sweat stays on the skin and doesn't evaporate. It then has very little cooling effect.

Secondly, hairs on the surface of the skin lie flat against the skin's surface. This happens because of the relaxation of tiny muscles called **hair erector muscles** attached to the base of each hair. In cold conditions, these contract and the hairs are pulled upright. The hairs trap a layer of air next to the skin, and since air is a poor conductor of heat, this acts as insulation. In warm conditions, the thinner layer of trapped air means that more heat will be lost. This is not very effective in humans, because the hairs over most of our body do not grow very large. It is very effective in hairy mammals like cats or dogs. The same principle is used by birds, which 'fluff out' their feathers in cold weather.

> Students often describe vasodilation incorrectly. They talk about the blood vessels 'moving nearer the surface of the skin'. They don't *move* at all, it's just that more blood flows through the surface vessels.

Lastly, there are tiny blood vessels called capillary loops in the dermis. Blood flows through these loops, radiating heat to the outside, and cooling the body down. If the body is too hot, small arteries (arterioles) leading to the capillary loops **dilate** (widen). This increases the blood flow to the skin's surface (Figure 8.16) and is called **vasodilation**.

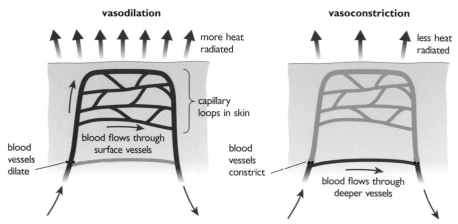

Figure 8.16 *Blood flow through the surface of the skin is controlled by vasodilation or vasoconstriction.*

In cold conditions, the opposite happens. The arterioles leading to the surface capillary loops **constrict** (become narrower) and blood flow to the surface of the skin is reduced, so that less heat is lost. This is called **vasoconstriction**. Vasoconstriction and vasodilation are brought about by tiny rings of muscles in the walls of the arterioles, called sphincter muscles, like the sphincters you met earlier in this chapter, at the outlet of the bladder.

There are other ways that the body can control heat loss and heat gain. In cold conditions, the body's **metabolism** speeds up, generating more heat. The liver, a large organ, can produce a lot of metabolic heat in this way. The hormone **adrenaline** stimulates the increase in metabolism (see Chapter 6). **Shivering** also takes place, where the muscles contract and relax rapidly. This also generates a large amount of heat.

Sweating, vasodilation and vasoconstriction, hair erection, shivering and changes to the metabolism, along with behavioural actions, work together to keep the body temperature to within a few tenths of a degree of the 'normal' 37 °C. If the difference is any bigger than this it shows that something is wrong. For instance, a temperature of 39 °C might be due to an illness.

End of Chapter Checklist

You should now be able to:

✓ understand that organisms are able to respond to changes in their environment

✓ recall that homeostasis is the maintenance of a constant internal environment

✓ recall that body maintenance of water content, temperature and composition of the blood are examples of homeostasis

✓ recall that excretion is the removal of metabolic waste, including urea, carbon dioxide and water

✓ recall that the lungs, kidneys and skin are organs of excretion

✓ recall that the kidney carries out the roles of excretion and osmoregulation

✓ recall the structure of the urinary system, including the kidneys, ureters, bladder and urethra

✓ recall the composition of urine, and understand how and why this may vary

✓ describe the structure and functions of a nephron, including the Bowman's capsule, glomerulus, convoluted tubules, loop of Henlé and collecting duct

✓ describe the process of ultrafiltration in the Bowman's capsule and the composition of the glomerular filtrate

✓ recall that water is reabsorbed into the blood from the collecting duct

✓ recall that selective reabsorption of glucose occurs at the proximal convoluted tubule

✓ describe the role of ADH in regulating the water content of the blood

✓ describe the value of kidney transplants and understand the problems of tissue matching to avoid rejection

✓ understand the use of kidney dialysis machines

✓ explain the importance of rehydration following loss of body fluids through vomiting and diarrhoea

✓ describe the oral rehydration method

✓ describe the role of the skin in temperature regulation

✓ understand the role of negative feedback with reference to temperature control and ADH secretion

Questions

1 Explain the meaning of the following terms:

   a) homeostasis

   b) excretion

   c) ultrafiltration

   d) selective reabsorption

   e) endotherm.

2 The diagram below shows a simple diagram of a nephron (kidney tubule).

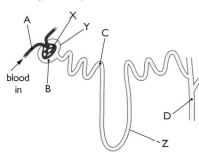

a) What are the names of the parts labelled X, Y and Z?

b) Four places in the nephron and its blood supply are labelled A, B and C and D. Which of the following substances are found at each of these four places?

    water   urea   protein   glucose   salt

3 The hormone ADH controls the amount of water removed from the blood by the kidneys. Write a short description of the action of ADH in a person who has lost a lot of water by sweating, but has been unable to replace this water by drinking. Explain how this is an example of negative feedback. (You need to write about half a page to answer this question fully.)

4 The bar chart shows the volume of urine collected from a person before and after drinking 1000 cm³ (1 dm³) of distilled water. The person's urine was collected immediately before the water was drunk and then at 30 minute intervals for four hours.

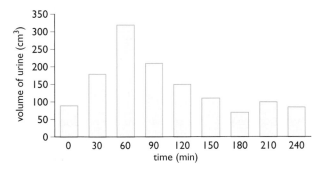

a) Describe how the output of urine changed during the course of the experiment.

b) Explain the difference in urine produced at 60 minutes and at 90 minutes.

c) The same experiment was repeated with the person sitting in a very hot room. How would you expect the volume of urine collected to differ from the first experiment? Explain your answer.

d) Between 90 and 120 minutes, the person produced 150 cm³ of urine. If the rate of filtration at the glomeruli during this time was 125 cm³ per minute, calculate the percentage of filtrate reabsorbed by the kidney tubules.

5 Working on a computer, construct a table to show the changes that take place when a person is put in a hot or cold environment. Your table should have three columns.

Changes taking place	Hot environment	Cold environment
sweating		
blood flow through capillary loops		vasoconstriction decreases blood flow through surface capillaries so that less heat is radiated from the skin
hairs in skin		
shivering		
metabolism		

6 Humans are able to maintain a constant body temperature, which is usually higher than that of their surroundings.

a) Explain the advantage in maintaining a constant high body temperature.

b) The temperature of the blood is constantly monitored by the brain. If it detects a drop in blood temperature, the following things happen: the arterioles leading to the skin capillaries constrict, less sweat is formed and shivering begins.

    i) Explain how each response helps the body to keep warm.

    ii) Explain how the structure of arterioles allows them to constrict.

c) When the weather is hot, we produce less urine.

    i) What is the name of the hormone that controls the amount of urine produced by the body?

    ii) Explain why the body produces less urine in hot weather.

    iii) Explain how the hormone in i) works in the kidney to produce less urine.

Chapter 9: Reproduction

Reproduction allows humans to pass on their genes. This occurs at fertilisation, through the fusion of special sex cells or gametes – the sperm and egg (ovum). In this chapter we look at human reproduction, from production of gametes to birth of the baby, growth and development, to adulthood.

The plural of sperm is sperm. The plural of ovum is ova.

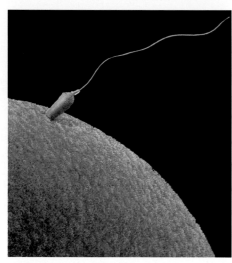

Figure 9.1 *A sperm fertilising an ovum.*

A gene is a section of DNA that determines a particular characteristic or feature. Genes are found in the nucleus of a cell on the chromosomes.

In sexual reproduction, specialised **sex cells** called **gametes** are produced. There are usually two types, a mobile male gamete (a **sperm**) and a stationary female gamete (an **ovum**).

The sperm must move to the ovum and fuse (join) with it. This is called fertilisation (Figure 9.1). The single cell formed by fertilisation is called a zygote. This cell will divide many times by mitosis to form all the cells of the new animal.

Sexual reproduction produces children who are different from either parent – in other words, they show **genetic variation**. This is partly because a child gains their genes from both parents, but also because new combinations of genes are produced by the cell division that forms gametes.

There are four key stages in sexual reproduction.

1. Gametes (sperm and ova) are produced.

2. The male gamete (sperm) is transferred to the female gamete (ovum).

3. Fertilisation must occur – the sperm fuses with the ovum.

4. The zygote formed develops into a new individual.

Production of gametes

Sperm are produced in the male sex organs – the **testes**. Ova are produced in the female sex organs – the **ovaries**. Both are produced when cells inside these organs divide. These cells do not divide by mitosis but by **meiosis** (see Chapter 11). Meiosis produces cells that are not genetically identical and have only half the number of chromosomes as the original cell. Figure 9.2 shows how a cell with just four chromosomes divides by meiosis. These four chromosomes are in two pairs called **homologous pairs**. Homologous pairs of chromosomes carry the same genes in the same sequence.

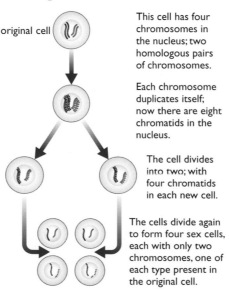

original cell

This cell has four chromosomes in the nucleus; two homologous pairs of chromosomes.

Each chromosome duplicates itself; now there are eight chromatids in the nucleus.

The cell divides into two; with four chromatids in each new cell.

The cells divide again to form four sex cells, each with only two chromosomes, one of each type present in the original cell.

Figure 9.2 *The stages of meiosis.*

In meiosis the cell divides twice, rather than just once as in mitosis. Also, because each of the sex cells formed only receives one chromosome from each original pair, they only have *half* the original number of chromosomes. They are **haploid** cells. Table 9.1 compares mitosis and meiosis.

Feature	Mitosis	Meiosis
number of cell divisions	1	2
number of cells formed	2	4
number of chromosomes in cells formed	same as original cell (diploid)	half the number of original cell (haploid)
type of cells formed	body cells	sex cells
genetic variation in cells formed	none	variation

Table 9.1: *Mitosis and meiosis compared.*

Transfer of the sperm to the ovum

Sperm are specialised for swimming (Figure 9.3). They have a tail-like flagellum that is able to propel them through fluids, and mitochondria in the mid-piece to supply ATP for energy. In humans, during sexual intercourse the male ejaculates sperm into the female reproductive system in a special fluid called semen. Semen provides nutrients for the sperm, and allows them to swim to meet the ovum.

Fertilisation

Once the sperm has reached the ovum, its nucleus must enter the ovum and fuse with the ovum nucleus. As each gamete has only half the normal number of chromosomes, the zygote formed by fertilisation will have the full number of chromosomes. In humans, sperm and ovum each have only 23 chromosomes. The zygote has 46 chromosomes, like all other cells in the body. Figure 9.4 shows the main stages in fertilisation.

Cells that have the full number of chromosomes in homologous pairs are called **diploid** cells. Cells that only have half the normal number of chromosomes are called **haploid** cells.

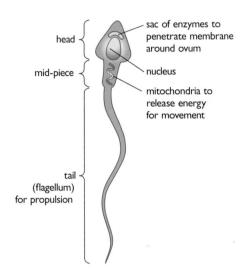

Figure 9.3 *The structure of a sperm.*

Red blood cells are exceptions. They have no nucleus, so have no chromosomes.

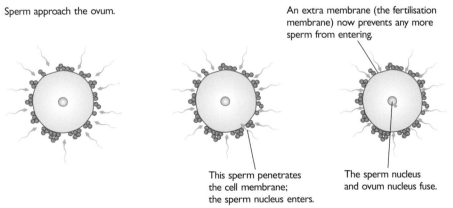

Figure 9.4 *The main stages in fertilisation.*

Fertilisation does more than just restore the diploid chromosome number, it provides an additional source of genetic variation. The sperm and ova are all genetically different because they are formed by meiosis. Therefore, each time fertilisation takes place, it brings together a different combination of genes.

Development of the zygote

Each zygote that is formed must divide to produce all the cells that will make up the adult. All these cells must have the full number of chromosomes, so the zygote divides repeatedly by mitosis. Figure 9.5 shows the importance of meiosis, mitosis and fertilisation in the human life cycle.

However, mitosis is not the only process involved in development, otherwise all that would be produced would be a ball of cells. During the process, cells move around and different shaped structures are formed. Also, different cells specialise to become bone cells, nerve cells, muscle cells, and so on.

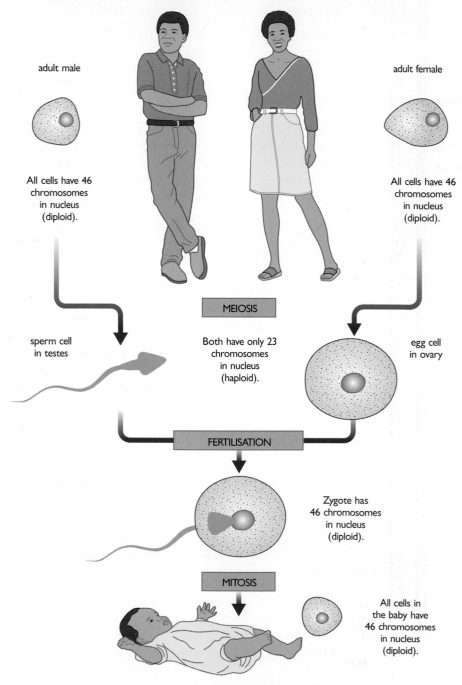

adult male

All cells have 46 chromosomes in nucleus (diploid).

adult female

All cells have 46 chromosomes in nucleus (diploid).

MEIOSIS

sperm cell in testes

Both have only 23 chromosomes in nucleus (haploid).

egg cell in ovary

FERTILISATION

Zygote has 46 chromosomes in nucleus (diploid).

MITOSIS

All cells in the baby have 46 chromosomes in nucleus (diploid).

Figure 9.5 *The importance of meiosis, mitosis and fertilisation in the human life cycle.*

The human reproductive systems

Figures 9.6 and 9.7 show the structures of the female and male reproductive systems.

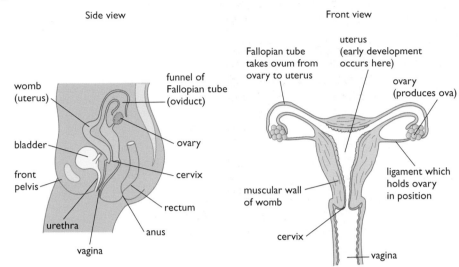

Figure 9.6 *The human female reproductive system.*

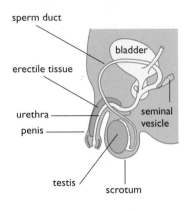

Figure 9.7 *The human male reproductive system.*

The sperm are produced in the testes by meiosis. During sexual intercourse, they pass along the sperm duct and are mixed with a fluid from the seminal vesicles. This mixture, called **semen**, is ejaculated into the vagina of the female. The sperm then begin to swim towards the Fallopian tubes.

One ovum is released into a Fallopian tube each month from an ovary. If an ovum is present in the Fallopian tubes, then it may be fertilised by sperm introduced during intercourse. The zygote formed will begin to develop into an **embryo** which will **implant** in the lining of the uterus. Here, the embryo will develop a **placenta**, which will allow the embryo to obtain materials such as oxygen and nutrients from the mother's blood. It also allows the embryo to get rid of waste products such as urea and carbon dioxide, as well as anchoring the embryo in the uterus. The placenta secretes female hormones, in particular progesterone, that maintain the pregnancy and prevent the embryo from aborting. Figure 9.8 shows the structure and position of the placenta.

During pregnancy, a membrane called the **amnion** encloses the developing embryo. The amnion secretes a fluid called **amniotic fluid**, which protects the developing embryo against jolts and bumps. As the embryo develops, it becomes more and more complex. When it becomes recognisably human, we no longer call it an embryo but a **fetus**. At the end of 9 months of development, there just isn't any room left for the fetus to grow and it sends a hormonal 'signal' to the mother to initiate birth. This is called 'going into labour'. Figure 9.8 also shows the position of a human fetus just before birth.

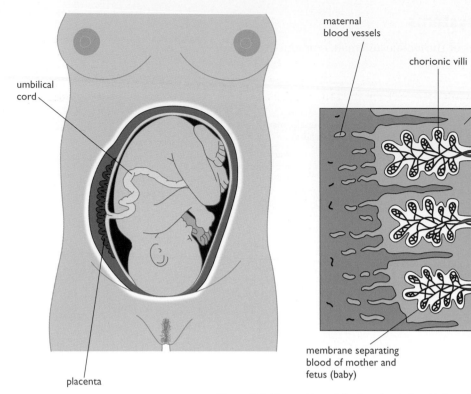

maternal blood vessels

chorionic villi

pool of mother's blood

umbilical cord

umbilical artery (carries deoxygenated blood containing waste products from fetus to placenta)

umbilical vein (carries oxygenated blood containing nutrients from placenta to fetus)

umbilical cord

placenta

membrane separating blood of mother and fetus (baby)

Figure 9.8 *The position of the fetus just before birth, and the structure of the placenta.*

Just before birth, the fetus takes up so much room that many of the mother's organs become displaced. The heart is pushed upwards and rotates so that the base points towards the left breast.

There are three stages to the birth of a child.

1. **Dilation of the cervix.** The cervix gets wider to allow the baby to pass through. The muscles of the uterus contract quite strongly and rupture the amnion, allowing the amniotic fluid to escape. This is called the breaking of the waters.

2. **Delivery of the baby.** Strong contractions of the muscles of the uterus push the baby head first through the cervix and vagina to the outside world.

3. **Delivery of the afterbirth.** After the baby has been born, the uterus continues to contract and pushes the placenta out, together with the membranes that surrounded the baby. These are known as the afterbirth.

Figure 9.9 shows the stages of birth.

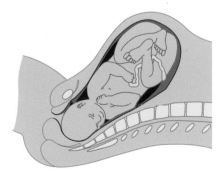

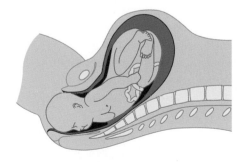

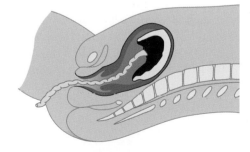

1 The baby's head pushes against the cervix; the mucous plug dislodges and the waters break.

2 The uterus contracts to push the baby out through the vagina.

3 The placenta becomes detached from the wall of the uterus and is expelled through the vagina as the afterbirth.

Figure 9.9 *The stages of birth.*

Breastfeeding

Within hours of its birth, a baby will be fed by its mother on milk from her **mammary glands**. Babies continue to feed on milk for several months, until they start to eat semi-solid foods. During pregnancy, the mother's breasts increase in size as the mammary glands grow new glandular tissue. They start to produce milk under the influence of a hormone called **prolactin** from the mother's pituitary gland. Another hormone from the pituitary called **oxytocin** stimulates the release of milk from the mammary glands.

While she is breastfeeding, the balance of the mother's diet has to change to accommodate the needs of milk production. Her energy intake from food needs to increase by about 25%, together with increases in protein, calcium and vitamins in her diet.

There are several advantages to breastfeeding. Breast milk is the perfect food for healthy growth of the baby. It also contains antibodies, which help to protect the baby against infectious diseases. Breastfeeding also helps form an emotional bond between the mother and her offspring. While it is certainly true that breastfeeding is best for both mother and baby, sometimes, for medical or other reasons, a mother cannot breastfeed. The alternative is for the baby to be fed on a special artificial milk preparation, called 'formula' milk.

When the baby sucks at the mother's nipple, a nervous reflex feeds back to the pituitary, stimulating it to release more oxytocin, in turn causing release of more milk. This is an example of a **positive feedback** system. You have seen several examples of negative feedback – positive feedback is much less common in the body.

Hormones controlling reproduction

Most animals are unable to reproduce when they are young. We say that they are sexually immature. When a baby is born, it is recognisable as a boy or girl by its sex organs.

The presence of male or female sex organs is known as the primary sex characteristics. During their teens, changes happen to boys and girls that lead to sexual maturity. These changes are controlled by hormones, and the time when they happen is called **puberty**. Puberty involves two developments. The first is that the sex cells (eggs and sperm) start to be produced. The second is that the bodies of both sexes adapt to allow reproduction to take place. These events are triggered by hormones released by the pituitary gland (see Table 6.2, page 88) called **follicle stimulating hormone (FSH)** and **luteinising hormone (LH)**.

In boys, FSH stimulates sperm production, while LH instructs the testes to secrete the male sex hormone, **testosterone**. Testosterone controls the development of the male **secondary sexual characteristics**. These include growth of the penis and testes, growth of facial and body hair, muscle development and breaking of the voice (Table 9.2).

In girls, the pituitary hormones control the release of a female sex hormone called **oestrogen**, from the ovaries. Oestrogen produces the female secondary sex characteristics, such as breast development and the beginning of menstruation ('periods').

Sperm production is most efficient at a temperature of about 34 °C, just below the core body temperature (37 °C). This is why the testes are outside the body in the scrotum, where the temperature is a little lower.

In boys	In girls
sperm production starts	the menstrual cycle begins, and eggs are released by the ovaries every month
growth and development of male sexual organs	growth and development of female sexual organs
growth of armpit and pubic hair, and chest and facial hair (beard)	growth of armpit and pubic hair
increase in body mass; growth of muscles, e.g. chest	increase in body mass; development of 'rounded' shape to hips
voice breaks	voice deepens without sudden 'breaking'
sexual 'drive' develops	sexual 'drive' develops
	breasts develop

Table 9.2: *Changes at puberty.*

The age when puberty takes place can vary a lot, but it is usually between about 11 and 14 years in girls and 13 and 16 years in boys. It takes several years for puberty to be completed. Some of the most complex changes take place in girls, with the start of menstruation.

Hormones and the menstrual cycle

'Menstrual' means 'monthly', and in most women the cycle takes about a month, although it can vary from as little as 2 weeks to as long as 6 weeks (Figures 9.10 and 9.11). In the middle of the cycle is an event called **ovulation**, which is the release of a mature egg cell, or **ovum**.

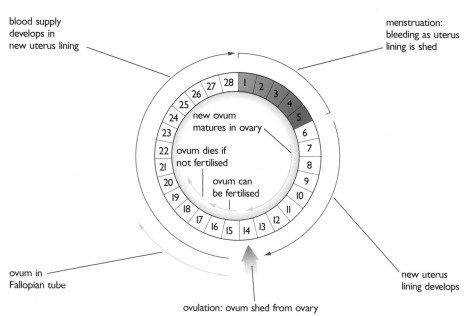

Figure 9.10 *The menstrual cycle.*

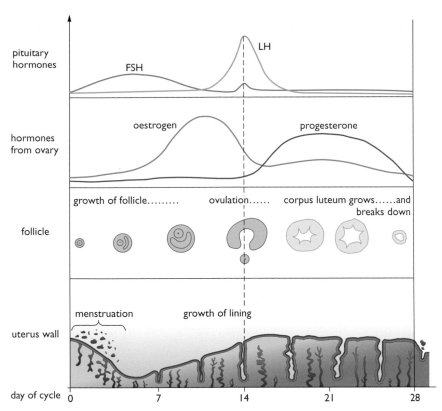

Figure 9.11 *Changes taking place during the menstrual cycle.*

One function of the cycle is to control the development of the lining of the uterus (womb), so that if the ovum is fertilised, the lining will be ready to receive the fertilised egg. If the egg is not fertilised, the lining of the uterus is lost from the woman's body as the flow of menstrual blood and cells of the lining, called a **period**.

A cycle is a continuous process, so it doesn't really have a beginning, but the first day of menstruation is usually called day 1.

Inside a woman's ovaries are hundreds of thousands of cells that could develop into mature eggs. Every month, one of these grows inside a ball of cells called a **follicle** (Figure 9.12). This is why the pituitary hormone which switches on the growth of the follicle is called 'follicle stimulating hormone'. At the middle of the cycle (about day 14) the follicle moves towards the edge of the ovary and the egg is released as the follicle bursts open. This is the moment of ovulation.

A small percentage of women are able to feel the exact moment that ovulation happens, as the egg bursts out of an ovary.

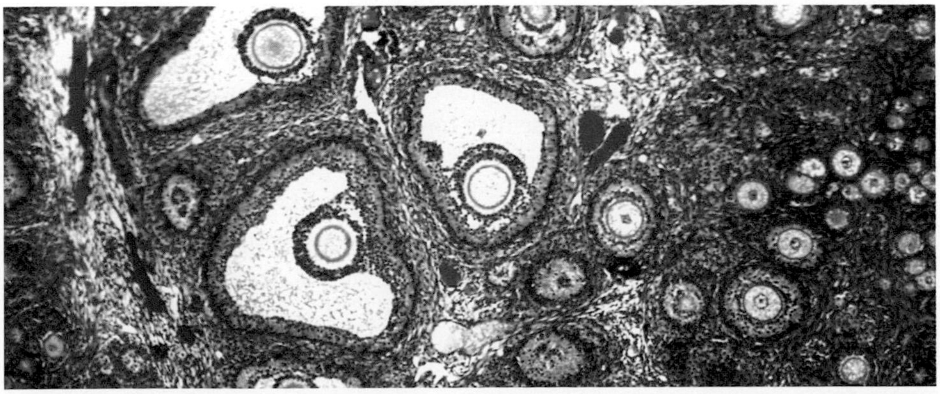

Figure 9.12 *Eggs developing inside the follicles of a human ovary*

'Corpus luteum' is Latin for 'yellow body'. A corpus luteum appears as a large yellow swelling in an ovary after the egg has been released. The growth of the corpus luteum is under the control of luteinising hormone (LH) from the pituitary.

While this is going on, the lining of the uterus has been repaired after menstruation, and has thickened. This change is brought about by the hormone oestrogen, which is secreted by the ovaries in response to FSH. Oestrogen also has another job. It slows down production of FSH, while stimulating secretion of LH. It is a peak of LH that causes ovulation.

After the egg has been released, it travels down the oviduct to the uterus. It is here in the oviduct that fertilisation may happen, if sexual intercourse has taken place. What's left of the follicle now forms a structure in the ovary called the **corpus luteum**. The corpus luteum makes another hormone called **progesterone**. Progesterone completes the development of the uterus lining, which thickens ready for the fertilised egg to sink into it and develop into an embryo. Progesterone also inhibits (prevents) the release of FSH and LH by the pituitary, stopping ovulation.

If the egg is not fertilised, the corpus luteum breaks down and stops making progesterone. The lining of the uterus is then shed through the woman's vagina, during menstruation. If, however, the egg is fertilised, the corpus luteum carries on making progesterone, the lining is not shed, and menstruation doesn't happen. The first sign that tells a woman she is pregnant is when her monthly periods stop. Later on in pregnancy, the **placenta** secretes progesterone, taking over the role of the corpus luteum.

Contraception

Contraception means avoiding conception – that is, avoiding becoming pregnant. Not having sexual intercourse is the most certain method of contraception, but many couples do want to have sex without the woman becoming pregnant. One way of doing this is the so-called 'natural' method. This uses knowledge of the menstrual cycle to avoid intercourse at times when sperm could possibly fertilise an ovum. Fertilisation is most likely to take place in the middle of the menstrual cycle, either side of the day of ovulation. Sperm and ova live for only a few days. Sperm released into the vagina 4 days before ovulation will have died by the time ovulation occurs. Two or 3 days after ovulation, an unfertilised ovum will have been lost through the vagina. By avoiding intercourse during this period, and limiting it to the rest of the menstrual cycle (the 'safe period'), a woman can usually avoid becoming pregnant (Figure 9.13).

For this method to be successful, the woman's menstrual cycle must be regular, and she must know when ovulation occurs in her cycle. The only disadvantage of this method is that it is very unreliable. There is usually a slight increase in a woman's body temperature around the time of ovulation, so she can take her temperature daily to try to judge when she is ovulating. Even if she does this, the 'safe period' method results in many unplanned pregnancies. However, some couples rely on the 'safe period' method, usually because they have moral or religious reasons for not using other means of contraception.

Another 'natural' method is where the man withdraws his penis from the woman's vagina before ejaculation. As with the 'safe period' method, this has the advantage that it does not involve using artificial contraception. But again, the 'withdrawal' method is unreliable – some sperm may be released before ejaculation, or the man may not withdraw in time.

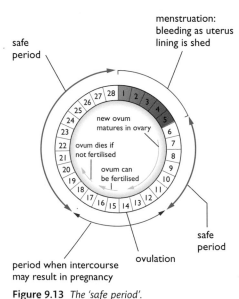

Figure 9.13 The 'safe period'.

There are four other categories of contraception:

- barrier methods
- intrauterine devices (IUDs)
- hormonal methods
- sterilisation.

Barrier methods use some kind of barrier to prevent sperm reaching the ovum. The **condom** is a sheath of very fine rubber that fits over the penis, and catches the semen before it enters the vagina (Figure 9.14). Another version, called the **femidom,** is similar, but inserted into the vagina before intercourse. Condoms and femidoms are easy to obtain and use, and have the added advantage that they are the only methods of contraception which give protection against sexually transmitted diseases.

Another barrier method is the **diaphragm** or **cap** (Figure 9.15). This is a dome-shaped piece of rubber that a woman inserts into her vagina before intercourse. The cap covers the cervix, preventing sperm from entering the uterus. A **spermicidal cream** is used with the cap as extra protection against pregnancy (spermicidal creams also increase the reliability of condoms). The cap is left in position for 6 hours after intercourse. Caps differ in size, and the woman needs a medical examination by a doctor or nurse to select the correct size of cap to use.

Intrauterine devices (IUD or coil) are small pieces of plastic or copper of various shapes. An IUD is inserted through the cervix into the uterus (Figure 9.16). The copper IUD works by preventing a fertilised egg from implanting in the lining of the uterus. Some IUDs contain the hormone progesterone, which thickens the mucus in the cervix, stopping sperm from getting through. One disadvantage of the IUD is that it has to be fitted by a doctor.

Hormonal methods mean taking the **oral contraceptive pill**. The **combined pill** contains a mixture of oestrogen and progesterone. These hormones prevent the production of FSH and LH from the pituitary gland (see page 127). This means that the follicles inside the ovary do not develop, and ovulation does not take place. Without an egg being released, a woman cannot become pregnant. She takes the pill for 21 days of the menstrual cycle (Figure 9.17) then stops taking it for the last 7 days. During this time she will have a period.

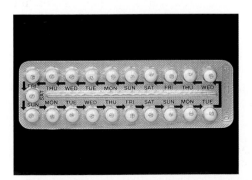

Figure 9.17 *Synthetic hormones can be used to prevent pregnancy.*

The **mini-pill** contains only progesterone. It works by causing a thickening of mucus in the cervix, which acts as a barrier to sperm.

Figure 9.14 *Condoms are easily obtained from clinics or pharmacists.*

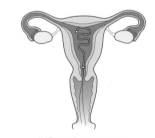

cap in position

the cap is lubricated with spermicidal cream before being placed over the cervix

Figure 9.15 *The contraceptive cap (diaphragm).*

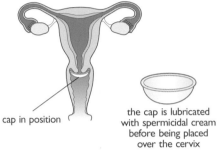

IUD in place in uterus

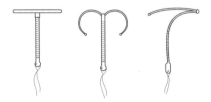

Figure 9.16 *Intrauterine devices.*

The combined pill has been linked to a number of health problems in women, the main one being an increased risk of blood clots (thromboses) forming in blood vessels, which can be fatal. The increased risk is only very slight for most women, but is higher if they are smokers. The mini-pill is not quite as reliable as the combined pill that contains oestrogen, but it does not increase the chances of blood clots developing.

Sterilisation is a surgical operation that can be carried out on men to prevent sperm passing to the penis, or on women to prevent eggs passing to the uterus. This makes them unable to have children (sterile). Sterilisation is a method that is normally only used by couples who have produced the number of children that they want, and do not want to use other methods of contraception to prevent further pregnancies. The operation is usually irreversible, so the decision to opt for sterilisation must be considered carefully.

Male sterilisation is called **vasectomy**. Under local anaesthetic, the sperm ducts are cut and tied so that no sperm can get through. After the operation, the man can still ejaculate but the semen will contain no sperm. In women, the Fallopian tubes are cut by a similar operation, called **tubal ligation**.

Table 9.3 compares the main methods of contraception in order of their rates of failure, and lists some of the advantages and disadvantages of each method.

Method	Failure rate (%)*		Some advantages and disadvantages
	Typical use	Careful use	
none – intercourse without contraception	85	85	
withdrawal	27	4	High failure rate.
'safe period'	25	9	High failure rate. Woman needs to have a regular cycle, and to keep records of the cycle.
diaphragm with spermicidal cream	16	6	Medical examination needed to select correct size. Not simple to use – must be inserted before intercourse and left in place for 6 hours afterwards.
condom	15	2	Easy to obtain and use. Gives protection against sexually transmitted diseases. May slip off during intercourse.
mini-pill	8	0.5	Low failure rate if used carefully. Must be taken every day, at the same time each day, to be effective.
combined pill	8	0.3	Low failure rate if used carefully. Must be taken every day. Links to some health problems.
intrauterine device	0.8	0.6	Must be fitted by a doctor. Can cause heavier periods.
tubal ligation	0.5	0.5	Very low failure rate. Operation usually not reversible.
vasectomy	0.1	0.1	Very low failure rate. Operation usually not reversible.

* The values for 'failure rate' are the number of pregnancies that result per 100 women per year. There are two sets of numbers. The 'careful use' column shows the failure rate when the contraceptives are used carefully, exactly as they should be. These values are from medical trials under controlled conditions. However, people are human, and make mistakes. The rates in the 'typical use' column are from surveys of couples who use the contraceptive methods normally, sometimes without taking the same degree of care as in the medical trials. This results in a higher rate of failure.

Table 9.3: *The main methods of contraception.*

Fertility treatment

Some couples have difficulty in achieving a pregnancy, even after years of trying. This medical problem is called **infertility**. There are many different reasons for infertility. For example the man may not produce enough sperm, the woman's eggs may not develop properly, or there may be a problem with achieving fertilisation.

Up to 40% of infertility problems are due to a low **sperm count**, so this is the first thing that doctors will check. The man's sperm count must be above 20 million sperm cells per cm^3 of semen, and the majority of the cells should show normal development.

If the sperm count is normal, the doctors will carry out tests on the woman to see if she is ovulating normally. If she is not, there are various treatments, depending on the nature of the problem.

There may be a physical problem, such as a blocked Fallopian tube or a cyst on the ovary. In these cases, surgery may be the answer.

Sometimes a woman does not produce enough follicle stimulating hormone (FSH) to start egg development in the ovary. This can be treated by giving her injections of FSH or other hormones, to stimulate ovulation. Care has to be taken, because too much FSH can produce multiple ovulations (many eggs are released at once). If they are all fertilised, this results in several embryos developing in the uterus, leading to 'multiple births' (Figure 9.18).

Although the children shown in Figure 9.18 were born healthy, multiple births may be premature (early). Even if the fetuses survive the full term of pregnancy (9 months), the mass of each baby is likely to be very low, and they may have health problems.

Sometimes **artificial insemination** (AI) is used. Healthy sperm are placed in a woman's uterus at the time of ovulation, to increase the chances of fertilisation. The woman's partner can donate the sperm, or sperm from another man can be used.

When all other methods of overcoming infertility problems have been tried, the last alternative is **in vitro** fertilisation (IVF). 'In vitro' is Latin for 'in glass', meaning fertilisation is carried out in a Petri dish or a similar container, outside the woman's body. Sperm is added to the egg, or a sperm can even be injected into an egg to fertilise it. The embryo's growth is monitored for a few days to check that all is well, then it is implanted into the woman's uterus to continue growth normally. This is now a common medical procedure that has produced many healthy 'test-tube babies'.

Even with a high sperm count, up to 40% of sperm are usually not viable. Some lack a head. Others have the appearance of a broken neck, with the head at 90° to the tail. Some even have two tails. Sperm that are deformed like this cannot swim or fertilise an egg.

Figure 9.18 *Hormones can also be used to help women become pregnant. These children were born to a mother who received treatment for infertility.*

Growth and development

A human develops from a fertilised egg into an embryo, then into a fetus. After birth, the baby grows into a child, then an adolescent, when it becomes sexually mature. Finally, he or she becomes an adult. This is the normal pattern of human development, the sequence through which the body changes during life. The instructions for development are carried in our genes. Most of the instructions are the same in everyone – well over 99% of our genes are the same in all other human beings. These genes control the basic growth pattern to produce a human body.

The remainder – a small fraction of the total genes – control the development of features that make each of us unique. As you have seen, this genetic variation comes from new assortments of genes being produced as a result of meiosis and fertilisation.

Genes act on growth and development in many different ways, for example, controlling the sex of the baby (Chapter 12, page 166). Sometimes genes produce their effects indirectly, by controlling the body's ability to make hormones, which in turn coordinate growth and development. For example, growth hormone, made by the pituitary gland, increases the rate of synthesis of proteins in muscles, and the growth of bone tissue. Thyroxine from the thyroid gland controls the metabolic rate (Chapter 6, page 89). Testosterone from the testes stimulates the development of the male secondary sex characteristics, and oestrogen from the ovary does the equivalent in girls. There are over a dozen different hormones that affect growth in some way.

Of course, growth and development also need a healthy balanced diet (Chapter 3). Lack of food will cause poor growth, and overeating will cause obesity.

The pattern of human growth

If you plot a graph of the height of a boy or girl against time, the graph isn't a straight line. Instead it shows growth 'spurts' – periods of rapid growth and development (Figure 9.19). The changes are easier to see if you plot the gain in height against the age of the child (Figure 9.20).

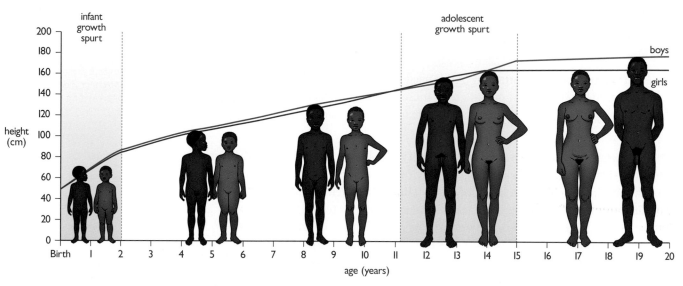

Figure 9.19 *Growth in boys and girls, showing changing rates of growth and growth spurts.*

A healthy child is always growing – the curve in Figure 9.20 is always greater than zero. But the *rate* of growth changes dramatically between birth and adolescence. Growth is fastest soon after birth. This is called the **infant growth spurt**. The rate then decreases rapidly until the child is about 4 years old. Between about 4 and 10 years the rate of growth still decreases, but more slowly. At puberty there is an increase in the rate of growth – the **adolescent growth spurt**. This happens rather earlier in girls compared with boys. Growth in both sexes has more or less finished by the late teens.

There is also a change in body proportions during growth. For example, a baby's head is very large in proportion to its body (Figure 9.21).

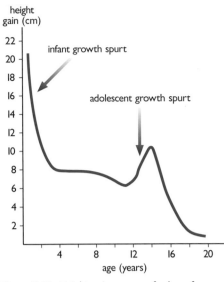

Figure 9.20 *Height gain per year for boys from birth to age 18 (the curve for girls is similar but with an adolescent spurt about 18 months earlier).*

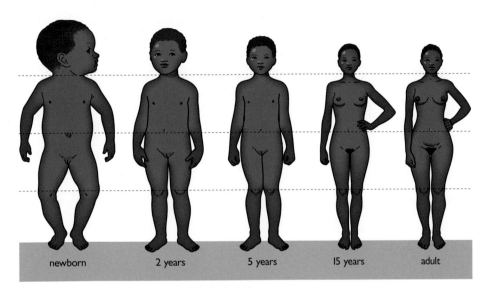

Figure 9.21 *Changes in proportions of the body with age.*

Different organs grow at different rates (Figure 9.22). Some organs like the kidney and liver keep pace with the rate of growth of the body as a whole. Others, such as the reproductive organs, grow much more slowly at first, and only catch up with the rest of the body after puberty. The brain and skull grow very rapidly early on, reaching 90% of their full size by the age of 6. the lymphatic system (see Chapter 4, page 60) also grows rapidly during the early years, but this rapid growth continues after the age of 6, until by about the age of 12 it reaches nearly double its final adult size, before decreasing again.

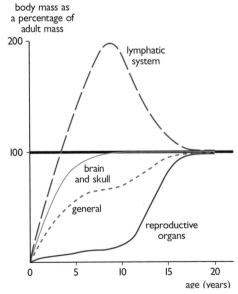

Figure 9.22 *Growth curves for different parts of the body compared with overall body growth.*

End of Chapter Checklist

Questions

1 The diagram shows a baby about to be born.

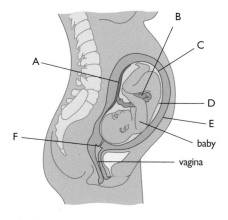

a) Name parts A to F on the diagram.

b) What is the function of A during pregnancy?

c) What must happen to D and E just before birth?

d) What must E and F do during birth?

2 a) The diagram shows the female reproductive system.

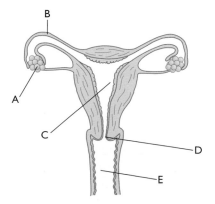

Which letter represents:

i) the site of production of oestrogen and progesterone

ii) the structure where fertilisation usually occurs

iii) the structure that must dilate when birth commences

iv) the structure that releases ova?

b) The graph shows the changes in the thickness of the lining of a woman's uterus over 100 days.

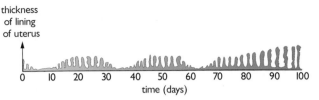

thickness of lining of uterus

time (days)

i) Name the hormone that causes the thickening of the uterine lining.

ii) Use the graph to determine the duration of this woman's menstrual cycle. Explain how you arrived at your answer.

iii) From the graph, deduce the approximate day on which fertilisation leading to pregnancy took place. Explain how you arrived at your answer.

iv) Why must the uterus lining remain thickened throughout pregnancy?

3 The number of sperm cells per cm³ of semen (the fluid containing sperm) is called the 'sperm count'. Some scientists believe that over the last 50 years, the sperm counts of adult male humans have decreased. They think that this is caused by a number of factors, including drinking water polluted with oestrogens and other chemicals. Carry out an Internet search to find out the evidence for this. Download information into a word processor and summarise your findings in no more than two sides of A4, including graphs.

4 The graph shows some of the changes taking place during the menstrual cycle.

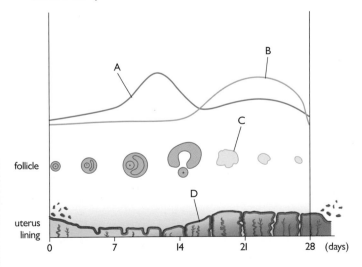

follicle

uterus lining

a) Identify the two hormones produced by the ovary, which are shown by the lines A and B on the graph.

b) Name the structure C.

c) What is the purpose of the thickening of the uterus lining at D?

d) When is sexual intercourse most likely to result in pregnancy, at day 6, 10, 13, 20 or 23?

e) Why is it important that the level of progesterone remains high in the blood of a woman during pregnancy? How does her body achieve this:

i) just after she becomes pregnant

ii) later on in pregnancy?

5 The diagram represents a typical menstrual cycle.

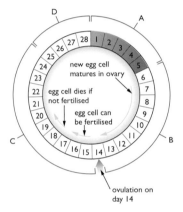

new egg cell matures in ovary

egg cell dies if not fertilised

egg cell can be fertilised

ovulation on day 14

a) Using evidence from the diagram, answer the following questions.

During which of the stages A, B, C or D does:

i) the level of the hormone oestrogen increase in the blood

ii) the level of the hormone progesterone increase in the blood

iii) the uterine lining become more vascular

iv) the levels of oestrogen and progesterone in the blood fall

v) the uterine lining begin to break down?

b) Explain how knowledge of the menstrual cycle can be used to avoid pregnancy.

6   Cells can divide by mitosis or by meiosis. Human cells contain
    46 chromosomes. The graphs show the changes in the
    number of chromosomes per cell as two different human
    cells undergo cell division.

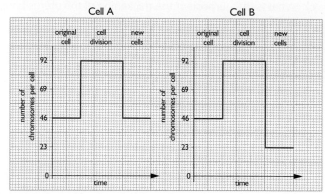

a) Which of the two cells, A or B, is dividing by meiosis?
    Explain how you arrived at your answer.

b) Explain the importance of meiosis, mitosis and fertilisation
    in maintaining the human chromosome number constant
    at 46 chromosomes per cell, generation after generation.

c) Give three differences between mitosis and meiosis.

7   a) Briefly explain how each of the following methods of
        contraception works:

    i)   the diaphragm (cap)

    ii) the intrauterine device (IUD)

    iii) the contraceptive pill (combined pill).

    b) What are the advantages and disadvantages of using
        the following methods of contraception:

    i)   the condom

    ii) the 'safe period'.

Chapter 10: Chromosomes, Genes and DNA

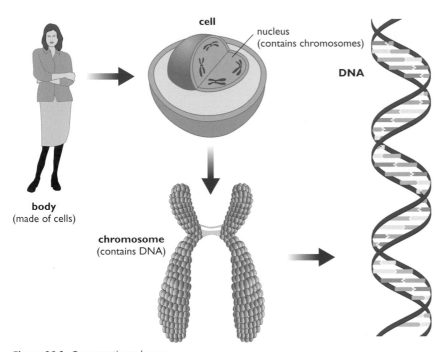

Figure 10.1 *Our genetic make-up.*

> *This chapter looks at the structure and organisation of genetic material, namely chromosomes, genes and DNA.*

> DNA is short for deoxyribonucleic acid. It gets its 'deoxyribo' name from the sugar in the DNA molecule. This is deoxyribose – a sugar containing five carbon atoms.

The chemical that is the basis of inheritance in nearly all organisms is **DNA**. DNA is usually found in the nucleus of a cell, in the **chromosomes** (Figure 10.1). A small section of DNA that determines a particular feature is called a **gene**. Genes determine features by instructing cells to produce particular proteins which then lead to the development of the feature. So a gene can also be described as a section of DNA that codes for a particular protein.

DNA can replicate (make an exact copy of) itself. When a cell divides by mitosis (see Chapter 11), each new cell receives exactly the same type and amount of DNA. The cells formed are **genetically identical**.

Figure 10.2 *(a) Watson and Crick with their double-helix model.*

The structure of DNA

Who discovered it?

James Watson and Francis Crick, working at Cambridge University, discovered the structure of the DNA molecule in 1953 (Figure 10.2a). Both were awarded the Nobel prize in 1962 for their achievement. However, the story of the first discovery of the structure of DNA goes back much further. Watson and Crick were only able to propose the structure of DNA because of the work of others – Rosalind Franklin (Figure 10.2b) had been researching the structure of a number of substances using a technique called X-ray diffraction.

Watson and Crick were able to use her results, together with other material, to propose the now-familiar double helix structure for DNA. Rosalind Franklin died

(b) Rosalind Franklin (1920–1958).

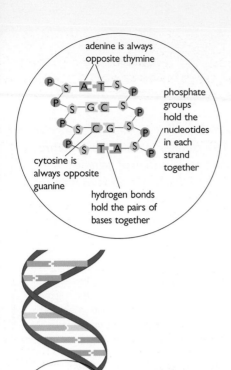

Figure 10.4 *Part of a molecule of DNA*

Key
P phosphate
S deoxyribose sugar
A adenine
T thymine
G guanine
C cytosine

Not only is the DNA code universal, but the actual DNA in different organisms is very similar. 98% of our DNA is the same as that of a chimpanzee; 50% of it is the same as that of a banana!

of cancer and so was unable to share in the award of the Nobel Prize (it cannot be awarded posthumously).

A molecule of DNA is made from two strands of **nucleotides**, making it a **polynucleotide**. Each nucleotide contains a nitrogenous base (adenine (A), thymine (T), cytosine (C) or guanine (G)), a sugar molecule and a phosphate group (Figure 10.3).

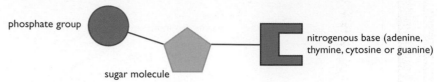

Figure 10.3 *The structure of a single nucleotide.*

Notice that, in the two strands (see Figure 10.4), nucleotides with adenine are always opposite nucleotides with thymine, and cytosine is always opposite guanine. Adenine and thymine are **complementary bases**, as are cytosine and guanine. Complementary bases always bind with each other and never with any other base. This is known as the **base-pairing rule**.

DNA is the only chemical that can replicate itself exactly. Because of this, it is able to pass genetic information from one generation to the next as a 'genetic code'.

The DNA code

Only one of the strands of a DNA molecule actually codes for the manufacture of proteins in a cell. This strand is called the **sense strand**. The other strand is called the **anti-sense** strand. The proteins manufactured can be **intracellular enzymes** (enzymes that control processes within the cell), **extracellular enzymes** (enzymes that are secreted from the cell to have their effect outside the cell), **structural proteins** (e.g. used to make hair, haemoglobin, muscles, cell membranes) or **hormones**.

Proteins are made of chains of amino acids. A sequence of *three* nucleotides in the sense strand of DNA codes for one amino acid. As the sugar and phosphate are the same in all nucleotides, it is actually the bases that code for the amino acid. For example, the base sequence TGT codes for the amino acid cysteine. Because three bases are needed to code for one amino acid, the DNA code is a **triplet code**. The sequence of bases that codes for *all* the amino acids in a protein is a gene (Figure 10.5).

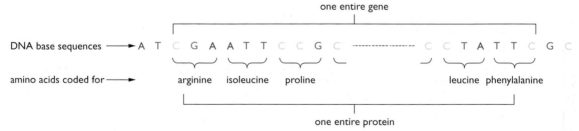

Figure 10.5 *The triplet code.*

The triplets of bases that code for individual amino acids are the same in all organisms. The base sequence TGT codes for the amino acid cysteine in humans, bacteria, bananas, monkfish, or in any other organism you can think of – the DNA code is a **universal code**.

DNA replication

When a cell is about to divide (see Mitosis, Chapter 11) it must first make an exact copy of each DNA molecule in the nucleus. This process is called **replication**. As a result, each cell formed receives exactly the same amount and type of DNA. Figure 10.6 summarises this process.

One consequence of the base-pairing rule is that, in each molecule of DNA, the amounts of adenine and thymine are equal, as are the amounts of cytosine and guanine.

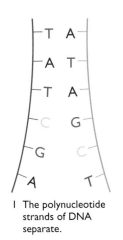

1 The polynucleotide strands of DNA separate.

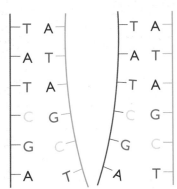

2 Each strand acts as a template for the formation of a new strand of DNA.

3 DNA polymerase assembles nucleotides into two new strands according to the base-pairing rule.

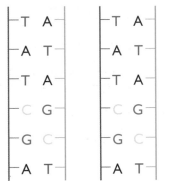

4 Two identical DNA molecules are formed – each contains a strand from the parent DNA and a new complementary strand.

Figure 10.6 *How DNA replicates itself.*

Gene mutations – when DNA makes mistakes

A **mutation** is a change in the DNA of a cell. It can happen in individual genes or in whole chromosomes. Sometimes, when DNA is replicating, mistakes are made and the wrong nucleotide is used. The result is a **gene mutation** and it can alter the sequence of the bases in a gene. In turn, this can lead to the gene coding for the wrong protein. There are several ways in which gene mutations can occur (Figure 10.7).

In **duplication**, Figure 10.7 (a), the nucleotide is inserted twice instead of once. Notice that the entire base sequence is altered – each triplet after the point where the mutation occurs is changed. The whole gene is different and will now code for an entirely different protein.

In **deletion**, Figure 10.7 (b), a nucleotide is missed out. Again, the entire base sequence is altered. Each triplet after the mutation is changed and the whole gene is different. Again, it will code for an entirely different protein.

In **substitution**, Figure 10.7 (c), a different nucleotide is used. The triplet of bases in which the mutation occurs is changed and it *may* code for a different amino acid. If it does, the structure of the protein molecule will be different. This may be enough to produce a significant alteration in the *functioning* of a protein or a total lack of function. However, the new triplet may not code for a different amino acid as most amino acids have more than one code.

In **inversions**, Figure 10.7 (d), the sequence of the bases in a triplet is reversed. The effects are similar to substitution. Only one triplet is affected and this may or may not result in a different amino acid and altered protein stucture.

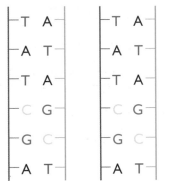

(a) ATT TCC GTT ATC
duplication here

ATT TTC CGT TAT C
extra T becomes first base of next triplet

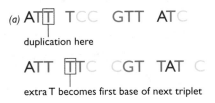

(b) ATT TCC GTT ATC
deletion here

ATT CCG TTA TC
replaced by first base of next triplet

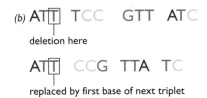

(c) ATT TCC GTT ATC
original base

ATG TCC GTT ATC
substituted base

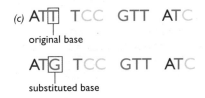

(d) ATT TCC GTT ATC
inversion here

ATT CCT GTT ATC

Figure 10.7 *Gene mutations (a) duplication, (b) deletion, (c) substitution, (d) inversion.*

Mutations that occur in body cells, such as those in the heart, intestines or skin, will only affect that particular cell. If they are very harmful, the cell will die and the mutation will be lost. If they are 'neutral' and do not affect the functioning of the cell in a major way, the cell may not die. If the cell then divides, a group of cells containing the mutant gene is formed. When the organism dies, however, the mutation is lost with it; it is not passed to the offspring. Only mutations in the sex cells or in the cells that divide to form sex cells can be passed on to the next generation. This is how genetic diseases begin. Some mutations are beneficial (see page 145).

It is important to realise that mutations are random events. In other words, they happen spontaneously – it is impossible to predict when a mutation will take place. However, the rate at which mutations occur is increased by a number of physical and chemical factors. These factors are called **mutagens**. Mutagens include:

- ionising radiation, such as ultraviolet (UV) light, X-rays and gamma rays

- many different chemicals, both natural and manmade.

Although natural mutagens are all around us, some activities lead to increased exposure to them. This causes a greater risk of cancer as a result of damage to the DNA of cells. For example, overexposure to UV light, through sunbathing or using sun beds, increases the likelihood of a person developing skin cancer. The smoke and tar from cigarettes contain over 40 known mutagenic chemicals, including powerful mutagens such as benzene, formaldehyde and N-nitrosamines.

> A mutation is a rare, random change in genetic material that can be inherited. Mutagens are agents that increase the rate of random mutations.

The structure of chromosomes

Each chromosome contains one double-stranded DNA molecule. The DNA is folded and coiled so that it can be packed into a small space. The DNA is coiled around proteins called **histones** (Figure 10.8).

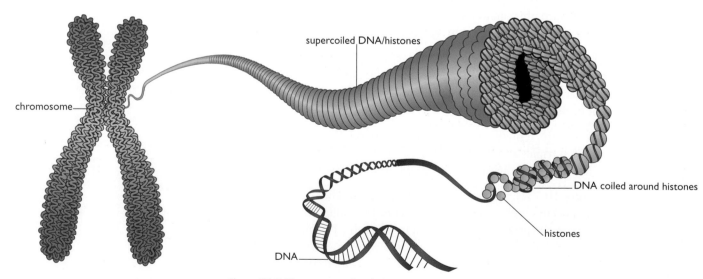

supercoiled DNA/histones

chromosome

DNA coiled around histones

histones

DNA

Figure 10.8 *The structure of a chromosome.*

Because a chromosome contains a particular DNA molecule, it will also contain the genes that make up that DNA molecule. Another chromosome will contain a different DNA molecule, and so will contain different genes.

How many chromosomes?

Nearly all human cells contain 46 chromosomes. The photographs in Figure 10.9 show the 46 chromosomes from the body cells of a human male and female.

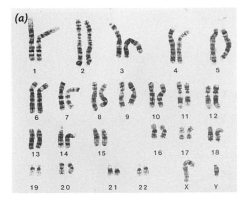
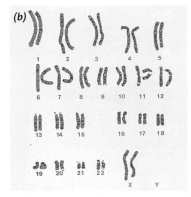

Figure 10.9 *Chromosomes of a human male (a) and female (b). A picture of all the chromosomes in a cell is called a karyotype.*

> Red blood cells have no nucleus, therefore no chromosomes. This gives them more room for carrying oxygen.

The chromosomes are not arranged like this in the cell. The original photograph has been cut up and chromosomes of the same size and shape 'paired up'. The cell from the male has 22 pairs of chromosomes and two that do not form a pair – the X and Y chromosomes. A body cell from a female has 23 matching pairs including a pair of X chromosomes.

> The X and the Y chromosomes are the **sex chromosomes**. They determine whether a person is male or female.

Pairs of matching chromosomes are called **homologous pairs**. They carry genes for the same features in the same sequence (Figure 10.10). Cells with chromosomes in pairs like this are **diploid** cells.

Not all human cells have 46 chromosomes. Red blood cells have no nucleus and so have none. Sex cells have only 23 – just half the number of other cells. They are formed by a cell division called **meiosis** (see Chapter 11). Each cell formed has one chromosome from each homologous pair, and one of the sex chromosomes. Cells with only half the normal diploid number of chromosomes, and therefore only half the DNA content of other cells, are **haploid** cells.

When two sex cells fuse in **fertilisation**, the two nuclei join to form a single diploid cell (a **zygote**). This cell has, once again, all its chromosomes in homologous pairs and two copies of every gene. It has the normal DNA content.

Genes and alleles

Genes are sections of DNA that control the production of proteins in a cell. Each protein contributes towards a particular body feature. Sometimes the feature is visible, such as eye colour or skin pigmentation. Sometimes the feature is not visible, such as the type of haemoglobin in red blood cells or the type of blood group antigen on the red blood cells.

a homologous pair of chromosomes

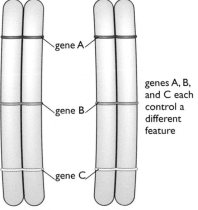

genes A, B, and C each control a different feature

Figure 10.10 *Both chromosomes in a homologous pair have the same sequence of genes.*

Some genes have more than one form. For example, the genes controlling several facial features have alternate forms, which result in alternate forms of the feature (Figure 10.11).

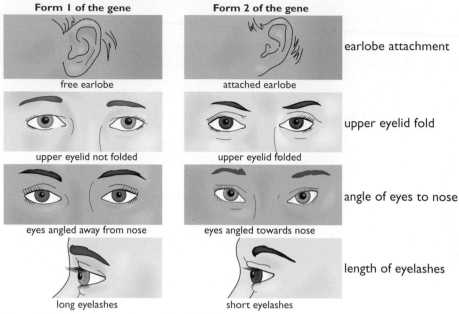

Form 1 of the gene	Form 2 of the gene	
free earlobe	attached earlobe	earlobe attachment
upper eyelid not folded	upper eyelid folded	upper eyelid fold
eyes angled away from nose	eyes angled towards nose	angle of eyes to nose
long eyelashes	short eyelashes	length of eyelashes

Figure 10.11 *The alternate forms of four facial features.*

The gene for earlobe attachment has the forms 'attached earlobe' and 'free earlobe'. These different forms of the gene are called **alleles**. Homologous chromosomes carry genes for the same features in the same sequence, but the alleles of the genes may not be the same (Figure 10.12). The DNA in the two chromosomes is not quite identical.

Each cell with two copies of a chromosome also has two copies of the genes on those chromosomes. Suppose that, for the gene controlling earlobe attachment, a person has one allele for attached earlobes and one for free earlobes. What happens? Is one ear free and the other attached? Are they both partly attached? Neither. In this case, both earlobes are free. The 'free' allele is **dominant** and 'switches off' the 'attached' allele, which is **recessive**. See Chapter 12 for more detail on how genes are inherited.

Chromosome mutations

When cells divide, they do not always divide properly. Bits of chromosomes can sometimes break off one chromosome and become attached to another. Sometimes one daughter cell ends up with both chromosomes of a homologous pair whilst the other has none. These 'mistakes' are called **chromosome mutations** and usually result in the death of the cells formed.

Sometimes sex cells do not form properly and they contain more (or fewer) chromosomes than normal. One relatively common chromosome mutation results in ova (female sex cells) containing two copies of chromosome 21. When an ovum like this is fertilised by a normal sperm, the zygote will have three copies of chromosome 21. This is called trisomy (three copies) of chromosome 21. Unlike

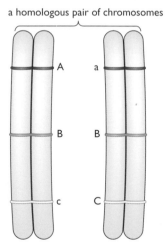

a homologous pair of chromosomes

Figure 10.12 *A and a, B and b, C and c are different alleles of the same gene. They control the same feature but code for different expressions of that feature.*

some other chromosome mutations, the effects of this mutation are usually non-fatal and the condition that results is **Down's syndrome** (Figure 10.13).

Down's syndrome children sometimes die in infancy, as heart and lung defects are relatively common. Those that survive have a near normal life span. Individuals with Down's syndrome can now live much more normal lives than was thought possible just 20 years ago. They require much care and attention during childhood, and particularly in adolescence, but, given this care, they can achieve good social and intellectual growth. Most importantly, they achieve personal self-sufficiency. Trisomy of chromosome 21 is more common in women over 40 years of age. As a result, they have more babies with Down's syndrome than younger women.

Figure 10.13 *This boy has Down's syndrome. His teacher is helping him to develop his full potential.*

Mutations and natural selection

As you have seen, many gene mutations are harmful, and cells that carry them will not usually survive. Some mutations are 'neutral', and if they arise in the gametes may be passed on without affecting the survival of the offspring. However, a few mutations can actually be beneficial to an organism. Beneficial mutations are the 'raw materials' that are ultimately the source of new inherited variation. A good example of this is seen in the phenomenon of bacterial resistance to antibiotics.

Alexander Fleming discovered the first antibiotic in 1929. It is a chemical produced by the mould *Penicillium*, and is called **penicillin**. Penicillin kills bacteria, and was first isolated and used to treat bacterial infections in the 1940s. Since then, other natural antibiotics have been discovered, and many more have been synthesised in laboratories. The use of antibiotics has increased dramatically, particularly over the past 20 years. We now almost expect to be given an antibiotic for even the most trivial of ailments. This can be dangerous, as it leads to the development of bacterial resistance to antibiotics (Figure 10.14).

Resistance starts when a random mutation gives a bacterium resistance to a particular antibiotic. In a situation where the antibiotic is widely used, the newly

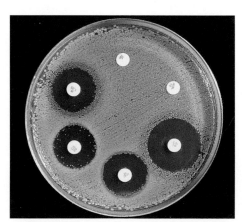

Figure 10.14 *This photo shows a colony of bacteria growing on a Petri dish of nutrient agar. The circular discs contain different antibiotics. Four of the discs have a clear area around them, where the bacteria have been killed by the antibiotics. The two discs at the top have had no effect on growth of the bacteria – they are resistant to the antibiotics in these discs.*

resistant bacterium has an advantage over non-resistant bacteria of the same type. The resistant strain of bacterium will survive and multiply in greater numbers than the non-resistant type. Bacteria reproduce very quickly – the generation time of a bacterium can be as short as 20 minutes. This means that there could be 72 generations in a single day. Very soon, a population of millions of resistant bacteria can be produced.

Bacterial resistance to antibiotics was first noticed in hospitals in the 1950s, and has grown to be a major problem today. It is a good example of **natural selection**. Natural selection was first proposed by the famous biologist Charles Darwin in the nineteenth century, as a mechanism that explains how **evolution** comes about. Darwin noticed that organisms in populations showed variation, and suggested that organisms that had advantages over others would be more likely to survive and reproduce. He called this idea 'survival of the fittest'. In the case of the bacteria, the resistant bacteria have a selective advantage over non-resistant bacteria – they are 'fitter'. In effect, the bacteria have evolved as a result of natural selection acting on the variation produced by mutation.

Doctors are now more reluctant to prescribe antibiotics. They know that by using them less, the bacteria with resistance will have less of an advantage and will not become widespread.

Natural selection is more obvious in bacteria, because they reproduce so rapidly that we can soon see its effects. But natural selection applies to all organisms, including humans. In Chapter 12 you will read about an example of natural selection in humans involving a disease called sickle cell anaemia.

A particularly worrying example of a resistant bacterium is MRSA. MRSA stands for methicillin-resistant *Staphylococcus aureus*. It is sometimes called a 'super bug' because it is resistant to many antibiotics, including methicillin (a type of penicillin). It is a particular problem in hospitals, where it is responsible for many difficult-to-treat infections.

You should now be able to:

✓ recall that the nucleus of a cell contains chromosomes on which genes are located

✓ recall that a gene is a section of a molecule of DNA

✓ describe the structure of a DNA molecule

✓ understand, in outline, how DNA acts as a genetic code

✓ understand, in outline, how DNA is replicated

✓ recall that mutation is a rare, random change in genetic material that can be inherited

✓ recall that many mutations are harmful, but some are neutral and a few are beneficial

✓ recall that the incidence of mutations can be increased by exposure to ionising radiation and some chemical mutagens

✓ recall that the diploid number of chromosomes in humans is 46 and the haploid number is 23 (see also Chapter 11)

✓ recall that genes exist in alternative forms called alleles, which can be dominant or recessive (see also Chapter 12)

✓ understand that mutations can increase in a population by natural selection (see also Chapter 12)

Questions

1   The diagram represents part of a molecule of DNA.

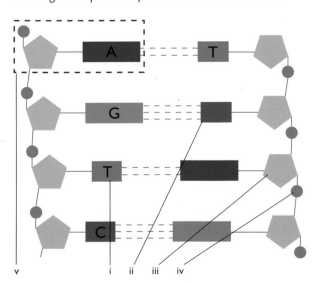

a) Name the parts labelled i, ii, iii, iv and v.

b) What parts did James Watson, Frances Crick and Rosalind Franklin play in discovering the structure of DNA?

c) Use the diagram to explain the base-pairing rule.

2   *a)* What is:

i)   a gene

ii) an allele?

b) Describe the structure of a chromosome.

c) How are the chromosomes in a woman's skin cells:

i)   similar to

ii) different from those in a man's skin cells?

3   DNA is the only molecule capable of replicating itself. Sometimes mutations occur during replication.

a) Describe how DNA replicates itself.

b) Explain how a single gene mutation can lead to the formation of a protein in which:

i)   many of the amino acids are different from those coded for by the non-mutated gene

ii) only one amino acid is different from those coded for by the non-mutated gene.

4 The graph shows the numbers and relative frequency of births of Down's syndrome babies in women aged between 20 and 50.

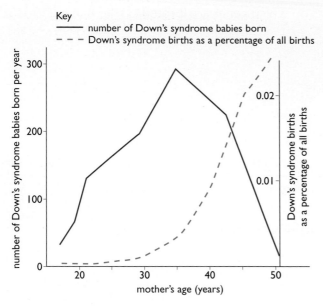

Key
—— number of Down's syndrome babies born
– – – Down's syndrome births as a percentage of all births

a) What is Down's syndrome?

b) How do the numbers of Down's syndrome births change with the age of the mother?

c) Suggest why the trend shown by the frequency of Down's syndrome births is different from that shown by the actual numbers.

Chapter 11: Cell Division

In most parts of the body, cells need to divide so that organisms can grow and replace worn out or damaged cells. The cells that are produced in this type of cell division should be exactly the same as the cells they are replacing. This is the most common form of cell division.

Only in the sex organs is cell division different. Here, some cells divide to produce gametes (sex cells), which contain only half the original number of chromosomes. This is so that when male and female gametes fuse together (fertilisation) the resulting cell (zygote) will contain the full complement of chromosomes and can then divide and grow into a new individual.

Human body cells have 46 chromosomes in 23 pairs called homologous pairs. Chromosomes in a homologous pair carry genes for the same features in the same sequence. They do not necessarily have the same *alleles* of every gene (see Chapter 10). These body cells are **diploid** cells – they have *two* copies of each chromosome. The sex cells, with 23 chromosomes (only one copy of each chromosome), are **haploid** cells.

There are two kind of cell division: **mitosis** and **meiosis**. When cells divide by mitosis, two cells are formed. These have the same number and type of chromosomes as the original cell. Mitosis forms all the cells in our bodies except the sex cells.

When cells divide by meiosis, four cells are formed. These have only half the number of chromosomes of the original cell. Meiosis forms sex cells.

There are two kinds of cell division in the human body. One type involves the process of mitosis, and is used to produce more body cells for growth. The other involves meiosis, and is used to produce gametes – eggs and sperm.

Meiosis is sometimes called **reduction division**. This is because it produces cells with only half the number of chromosomes of the original cell.

Mitosis

When a **parent cell** divides it produces **daughter cells**. Mitosis produces two daughter cells that are genetically identical to the parent cell – both daughter cells have the same number and type of chromosomes as the parent cell. To achieve this, the dividing cell must do two things.

- It must copy each chromosome before it divides. This involves the DNA replicating and more proteins being added to the structure. Each daughter cell will then be able to receive a copy of each chromosome (and each molecule of DNA) when the cell divides.

- It must divide in such a way that each daughter cell receives one copy of every chromosome. If it does not do this, both daughter cells will not contain all the genes.

These two processes are shown in Figure 11.1.

A number of distinct stages occur when a cell divides by mitosis. These are shown in Figure 11.2. Mitosis is easy to see in some actively dividing plant tissues. Figure 11.3 is a photograph of some cells from the root tip of an onion. Cells in this region of the root divide by mitosis to allow growth of the root.

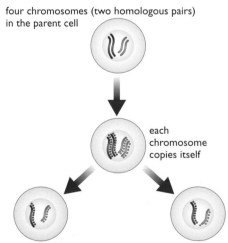

four chromosomes (two homologous pairs) in the parent cell

each chromosome copies itself

The cell divides into two; each new cell has a copy of each of the chromosomes.

Figure 11.1 *A summary of mitosis.*

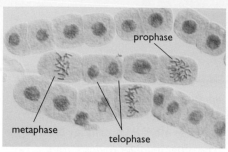

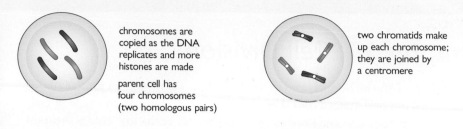

chromosomes are copied as the DNA replicates and more histones are made

parent cell has four chromosomes (two homologous pairs)

two chromatids make up each chromosome; they are joined by a centromere

Figure 11.3 *Photomicrograph of cells in the root tip of an onion. Cells showing some of the stages of mitosis are labelled.*

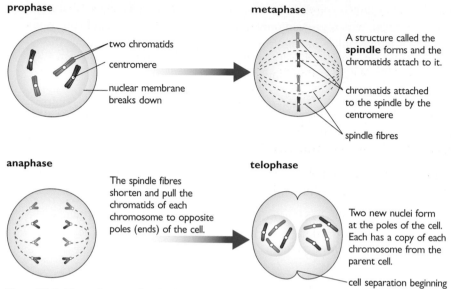

prophase

two chromatids

centromere

nuclear membrane breaks down

metaphase

A structure called the **spindle** forms and the chromatids attach to it.

chromatids attached to the spindle by the centromere

spindle fibres

anaphase

The spindle fibres shorten and pull the chromatids of each chromosome to opposite poles (ends) of the cell.

telophase

Two new nuclei form at the poles of the cell. Each has a copy of each chromosome from the parent cell.

cell separation beginning

Figure 11.2 *The main stages in mitosis.*

Each daughter cell formed by mitosis receives a copy of every chromosome, and therefore every gene, in the parent cell. Each daughter cell is genetically identical to the others. All the cells in our body (except the sex cells) are formed by mitosis from the zygote (single cell formed at fertilisation). They all, therefore, contain copies of all the chromosomes and genes of that zygote. They are all genetically identical.

Whenever cells need to be replaced in our bodies, cells divide by mitosis to make them. This happens more frequently in some regions than in others.

- The skin loses thousands of cells every time we touch something. This adds up to millions every day that need replacing. A layer of cells beneath the surface is constantly dividing to produce replacements.

- Cells are scraped off the lining of the gut as food passes along. Again, a layer of cells beneath the gut lining is constantly dividing to produce replacement cells.

- Cells in our spleen destroy worn out red blood cells at the rate of 100 000 000 000 per day! These are replaced by cells in the bone marrow dividing by mitosis. In addition, the bone marrow forms all our new white blood cells and platelets (Figure 11.4).

- Cancer cells divide by mitosis. The cells formed are exact copies of the parent cell, including the mutation in the genes that makes the cells divide uncontrollably.

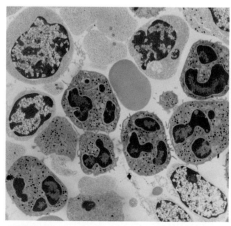

Figure 11.4 *Cells in bone marrow dividing to produce blood cells.*

Meiosis

Meiosis forms only sex cells. It is a more complex process than mitosis involving two cell divisions, but you don't need to know details of all the stages. Meiosis produces four cells that are haploid and not genetically identical. The dividing cell must do two things.

- It must copy each chromosome so that there is enough genetic material to be shared between the four daughter cells.

- It must divide twice, in such a way that each daughter cell receives just one chromosome from each homologous pair.

These processes are summarised in Figure 11.5. Figure 11.6 shows cells from a testis dividing by meiosis to produce sperm cells.

The sex cells formed by meiosis don't all have the same combinations of alleles – there is **genetic variation** in the cells. During the two cell divisions of meiosis, the chromosomes are divided between the two daughter cells independently of each other. There are only two 'rules'.

- During the first division of meiosis, one chromosome from each homologous pair goes into each daughter cell.

- During the second division of meiosis the chromosome separates into two parts. One part goes into each daughter cell. This allows for a lot of variation in the daughter cells (Figure 11.7).

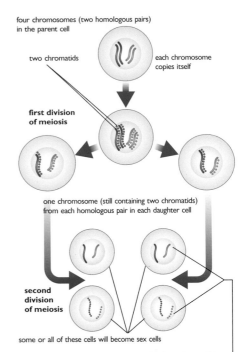

Figure 11.5 *A summary of meiosis.*

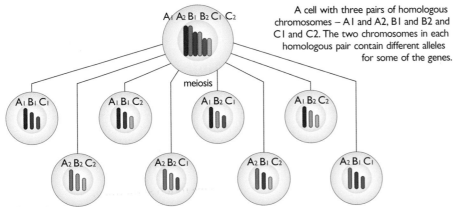

A cell with three pairs of homologous chromosomes – A1 and A2, B1 and B2 and C1 and C2. The two chromosomes in each homologous pair contain different alleles for some of the genes.

As a result of the two divisions of meiosis, each sex cell formed contains one chromosome from each homologous pair. This gives eight combinations. As A1 and A2 contain different alleles (as do B2 and B2, and C1 and C2) the eight possible sex cells will be genetically different.

Figure 11.7 *How meiosis produces variation.*

The features of mitosis and meiosis are show in Table 11.1.

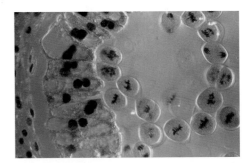

Figure 11.6 *Photomicrograph of cells from a testis dividing by meiosis.*

Feature of the process	Mitosis	Meiosis
do the chromosomes duplicate before division begins?	yes	yes
how many cell divisions are there?	one	two
how many cells are formed by the process?	two	four
are the cells formed haploid or diploid?	diploid	haploid
is there genetic variation in the cells formed?	no	yes

Table 11.1: *Comparison of meiosis and mitosis.*

There is a mathematical rule for predicting how many combinations of chromosomes there can be. The rule is:

number of possible combinations = 2^n

where n = number of *pairs* of chromosomes.

With two pairs of chromosomes, the number of possible combinations = $2^2 = 4$. With three pairs of chromosomes, the number of possible combinations = $2^3 = 8$. With the 23 pairs of chromosomes in human cells, the number of possible combinations = $2^{23} = 8\,388\,608$!

Sexual reproduction and variation

Sexual reproduction in any multicellular organism involves the fusion of two sex cells to form a zygote. The offspring from sexual reproduction vary genetically for a number of reasons. One reason is because of the huge variation in the sex cells. The other main reason is because of the random way in which fertilisation takes place. In humans, any one of the billions of sperm formed by a male during his life could, potentially, fertilise any one of the thousands of ova formed by a female.

This variation applies to both male and female sex cells. So, just using our 'low' estimate of about 8.5 million different types of human sex cells means that there can be 8.5 million different types of sperm and 8.5 million different types of ova. When fertilisation takes place, any sperm could fertilise any ovum. The number of possible combinations of chromosomes (and genes) in the zygote is 8.5 million × 8.5 million = 72 trillion! And remember, this is using our 'low' number!

This means that every individual is likely to be genetically unique. The only exceptions are **identical twins** (and identical triplets and quadruplets). Identical twins are formed from the *same* zygote – they are sometimes called **monozygotic twins**. When the zygote divides by mitosis, the two *genetically identical* cells formed do not 'stay together'. Instead, they separate and each cell behaves as though it were an individual zygote, dividing and developing into an embryo (Figure 11.8). Because they have developed from genetically identical cells (and, originally, from the same zygote), the embryos (and, later, the children and the adults they become) will be genetically identical.

Non-identical twins or **fraternal twins** develop from different zygotes and so are not genetically identical.

Asexual reproduction and cloning

When organisms reproduce asexually, there is no fusion of sex cells. A part of the organism grows and somehow breaks away from the parent organism. The cells it contains were formed by mitosis, so they contain exactly the same genes as the parent. Asexual reproduction produces offspring that are genetically identical to the parent. These are known as a **clone**. Asexual reproduction is common in plants, but less so in animals.

Sexual reproduction in plants results in the formation of seeds. Each seed contains an embryo, which results from the fusion of a pollen grain nucleus with an egg cell. This means that seeds from the same plant will vary genetically. Plant breeders have known about this variation for a long time. They realised that if a plant had some desirable feature, the best way to get more of that plant was not to collect its seeds, but to clone it in some way. Traditionally this has been done by growing plants from cuttings. A piece of the parent plant is removed and grown to produce more plants. Modern plant-breeding techniques do something similar, but from just a few cells of the original (Figure 11.9). This is called **micropropagation**, and is an important way that many crop plants for humans are grown.

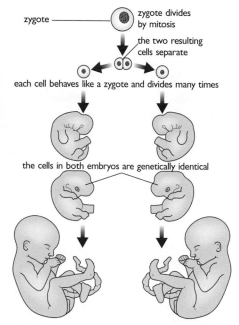

Figure 11.8 *How identical twins are formed.*

zygote

zygote divides by mitosis

the two resulting cells separate

each cell behaves like a zygote and divides many times

the cells in both embryos are genetically identical

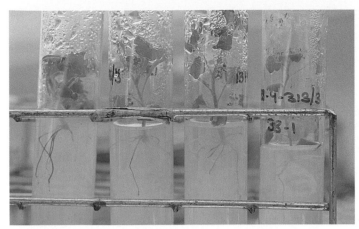

Figure 11.9 *Plantlets grown from a few cells of a parent plant by micropropagation.*

Some simple animals reproduce asexually, forming clones of the parent animal. For example, the tiny freshwater animal called *Hydra* (a relative of jellyfish) reproduces by forming 'buds' that develop into new individuals (Figure 11.10).

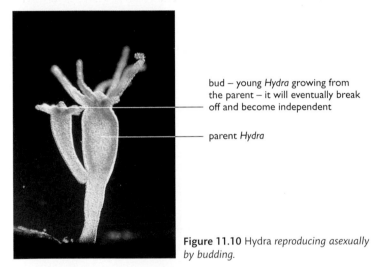

bud – young *Hydra* growing from the parent – it will eventually break off and become independent

parent *Hydra*

Figure 11.10 Hydra *reproducing asexually by budding.*

More complex animals such as mammals only reproduce sexually. However, artificial cloning in laboratories has been achieved in several species. The first cloned mammal was the famous Dolly the sheep. Dolly was produced by persuading one of her mother's ova (egg cells) to develop into a new individual without being fertilised by a sperm. The nucleus of the ovum was removed and 'replaced' with a cell taken from the udder of another sheep. The cell that was formed had the same genetic information as all the cells in the donor and so developed into an exact genetic copy. The stages in the procedure are shown in Figure 11.12. Figure 11.11 shows how an udder cell is inserted into an egg cell that has had its nucleus removed.

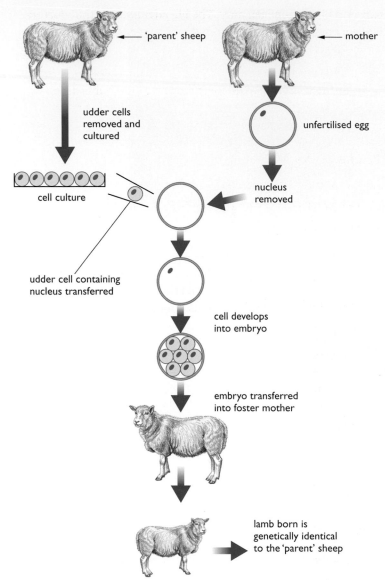

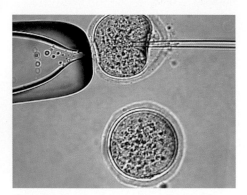

Figure 11.11 *Inserting an udder cell into an egg cell that has had its nucleus removed.*

Figure 11.12 *How Dolly was produced.*

Dolly was produced only after many unsuccessful attempts. Since then, the procedure has been repeated using other sheep, as well as rats, mice and pigs. Some of the animals produced are born deformed. Some do not survive to birth. Biologists believe these problems occur because the genes that are transferred to the egg are 'old genes'. These genes came from an animal that had already lived for several years, and from cells specialised to do things other than produce sex cells. It will take much more research to make the technique reliable.

Cloning a whole animal like this is called **reproductive cloning**. In theory, it would be possible to clone humans in this way. However, the idea of artificially cloning new humans from body cells is regarded by most people as ethically unsound, and research on human reproductive cloning is banned in many countries.

Research on another type of human cloning is allowed. This is the cloning of human body cells to provide cells or tissues for use in medicine. It is called therapeutic cloning. Therapeutic cloning could provide tissues and organs for

The only natural human clones are identical twins (see page 152).

transplants, to repair or replace damaged ones. The big advantage is that doctors should be able to clone a patient's own cells, so that any transplant would not be rejected by their immune system (see page 189).

In therapeutic cloning, a person's own DNA is used to grow an embryonic clone. However, instead of inserting this embryo into a surrogate mother, it is used to grow special cells called stem cells. These stem cells could be used to grow replacement organs such as heart, liver or skin. They could also be used to grow nerve cells to cure people who suffer from diseases of the nervous system, such as Alzheimer's or Parkinson's disease.

Therapeutic cloning uses some of the technology of reproductive cloning, but just to grow an embryo for its stem cells, not to produce a new individual:

- a nucleus is extracted from a body cell of the patient

- the nucleus is inserted into a donor egg that has had its nucleus removed

- the egg then divides to form an embryo containing stem cells

- stem cells are removed from the embryo

- a tissue or organ is grown from the stem cells to treat the patient.

There are still ethical issues associated with this research. Some people think it is wrong to use human embryos in this way, while others point out that the embryos that are used are at a very early stage in their development, and can't be thought of as a 'human life'. Certainly, in the future stem cell research and therapeutic cloning may offer ways of helping people with severe medical problems that at the moment are very difficult to treat by other means.

Genes and environment both produce variation

Identical twins have the same genes. They often grow up to look very alike, and often develop similar talents. However, they never look exactly the same. This is especially true if, for some reason, they grow up apart. The differences are due to differences in their environment affecting their appearance. In other words, both genes and environment have an effect on human variation.

A good example of this is adult human body mass (weight) or height. A person's growth is affected by many genes. There are genes that influence protein synthesis in muscles, bone development, production of hormones, etc. These will all have an effect on the growth of the body. But growth will take place only if the individual has access to a healthy balanced diet, so their weight will largely depend on environmental factors – in this case, availability of food.

Skin colour is also inherited (Figure 11.13) but it is also affected by the environment, in that exposure to the sun increases the amount of melanin in the skin, causing it to darken.

One of the most controversial issues in human variation is how much 'intelligence' is genetic or environmental. Does an intelligent child gain her genes for intelligence from her parents, or is much of it a result of the environment in which she grew up? This is sometimes called the nature/nurture argument. It has never been answered satisfactorily, and probably never will be, but it is certainly true that a child's intellectual development is affected by access to books and a good education, not to mention a healthy diet and good medical care.

What are stem cells? There are two types. **Embryonic stem cells** are found in the developing embryo, where they divide by mitosis and develop into all the specialised tissues. **Adult stem cells** are found in adult tissues. Here they act as a repair system for the body, retaining the ability to divide and differentiate into specialised tissues such as blood, skin or intestine, as part of the body's normal repair mechanisms.

Figure 11.13 *These children have inherited different combinations of genes for skin colour from their parents.*

End of Chapter Checklist

You should now be able to:

✓ understand that division of a diploid cell by mitosis produces two cells that contain identical sets of chromosomes

✓ understand a simple outline description of the four stages of mitosis

✓ recall that the diploid number of chromosomes in humans is 46 and the haploid number is 23

✓ understand that division of a cell by meiosis produces four cells, each with half the number of chromosomes, and that this results in the formation of genetically different haploid gametes

✓ understand that mitosis occurs during growth, repair, asexual reproduction and cloning

✓ understand that variation within a species can be genetic, environmental or a combination of both

Questions

1 Cells can divide by mitosis or by meiosis.

   a) Give one similarity and two differences between the two processes.

   b) Do cancer cells divide by mitosis or meiosis? Explain your answer.

   c) Why is meiosis sometimes called reduction division?

2 Some cells divide by mitosis, others divide by meiosis. For each of the following examples, say whether mitosis or meiosis is involved. In each case, give a reason for your answers.

   a) Cells in the testes dividing to form sperm.

   b) Cells in the lining of the small intestine dividing to replace cells that have been lost.

   c) Cells in the bone marrow dividing to form red blood cells and white blood cells.

   d) A zygote dividing to form an embryo.

3 Variation in organisms can be caused by the environment as well as by the genes they inherit. For each of the following examples, state whether the variation described is likely to be genetic, environmental or both. In each case, give a reason for your answers.

   a) Humans have brown, blue or green eyes.

   b) Half the human population is male, half is female.

   c) Cuttings of hydrangea plants grown in soils with different pH values develop flowers with slightly different colours.

   d) People in some families are more at risk of heart disease than people in other families. However, not every member of the 'high risk' families has a heart attack and some members of the 'low risk' families do.

4 In an investigation into mitosis, the distance between a chromosome and the pole (end) of a cell was measured. The graph shows the result of the investigation.

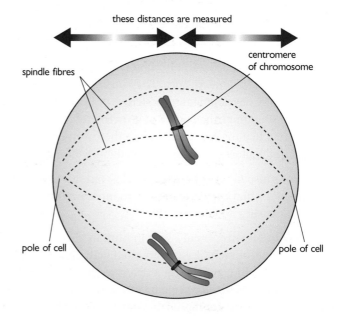

these distances are measured

centromere of chromosome

spindle fibres

pole of cell

pole of cell

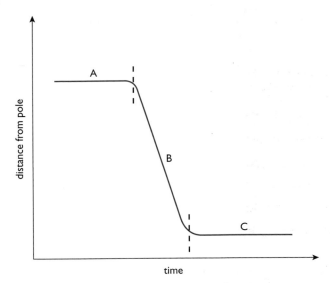

a) Describe two events that occur during stage A.

b) Explain what is happening during stage B.

c) Describe two events that occur during stage C.

5 Diagrams A–F show an animal cell during cell division.

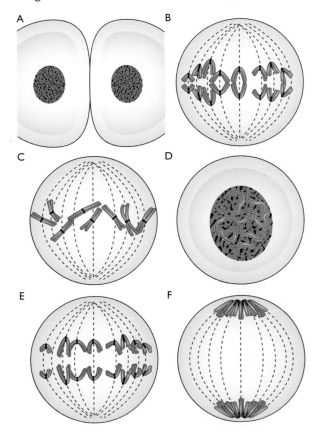

a) Put the pictures in the correct order.

b) Is the cell going through mitosis or meiosis? Explain your answer.

c) This cell has eight chromosomes which is its diploid number. How many chromosomes would a diploid human cell have?

d) Describe two differences between mitosis and meiosis.

6 The following flowchart shows how Dolly the sheep was cloned.

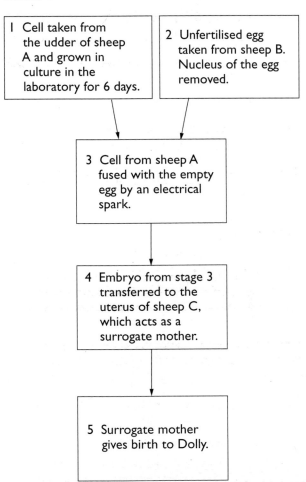

a) Where did scientists get the DNA to put into the unfertilised egg from sheep B?

b) How does the nucleus removed from an egg differ from the nucleus of an embryo?

c) Dolly is genetically identical to another sheep in the flowchart. Which one?

d) Give two ways in which this method is different from the normal method of reproduction in sheep.

e) Suggest two advantages of producing animal clones.

Chapter 12: Genes and Inheritance

This chapter deals with the rules of inheritance, which were first worked out by Gregor Mendel from the results of his experiments on pea plants. In his honour, they are called the laws of Mendelian genetics. Mendel's laws don't apply just to plants, but to nearly all organisms, including humans. Many human diseases are caused by faulty genes, so a knowledge of Mendelian genetics is important for understanding inherited diseases.

Genes are sections of DNA that determine a particular feature (see Chapter 10) by instructing cells to produce particular proteins. As the DNA is part of a chromosome, we can also define a gene as 'part of a chromosome that determines a particular feature'. The ground-breaking research that uncovered the basic rules of how features are inherited was carried out by Gregor Mendel and published in 1865.

Gregor Mendel

Gregor Mendel (Figure 12.1) was a monk who lived in a monastery in Brno in what is now the Czech Republic. He became interested in inheritance and his first attempts at controlled breeding experiments were with mice. This was not well received in the monastery and he was advised to use pea plants instead. As a result of the experiments with pea plants, he was able to formulate the basic laws of inheritance.

Mendel established that, for each feature he studied:

- a 'heritable unit' (we now call it a **gene**) is passed from one generation to the next
- the heritable unit (gene) can have alternate forms (we now call these different forms **alleles**)
- each individual must have two alternate forms (alleles) per feature
- the sex cells only have one of the alternate forms (allele) per feature
- one allele can be dominant over the other.

Mendel was able to use his ideas to predict outcomes from breeding certain types of pea plant and then test his predictions by experiment. Mendel published his results and ideas in 1865 but very few people took any notice.

At that time, biologists had little knowledge of chromosomes and cell division, so Mendel's ideas had no physical basis. Also, biology then was very much a descriptive science and biologists of the day were not interested in the mathematical treatment of results. Mendel's work went against the ideas of the time that inheritance resulted from some kind of blending of features. The idea of a distinct 'heritable unit' just did not fit in.

It was not until 1900 that other biologists working on inheritance rediscovered Mendel's work and recognised its importance. In 1903, the connection between Mendel's suggested behaviour of genes and the behaviour of chromosomes in meiosis was noticed. The science of genetics was well and truly born.

Figure 12.1 *Gregor Mendel (1822–1884).*

Mendel's experiments on inheritance

Mendel noticed that many of the features of pea plants had just two alternate forms. For example, plants were either tall or dwarf, they either had purple or white flowers; they produced yellow seeds or green seeds. There were no intermediate forms, no pale purple flowers or green/yellow seeds or intermediate height plants. Figure 12.2 shows some of the contrasting features of pea plants that Mendel used in his breeding experiments.

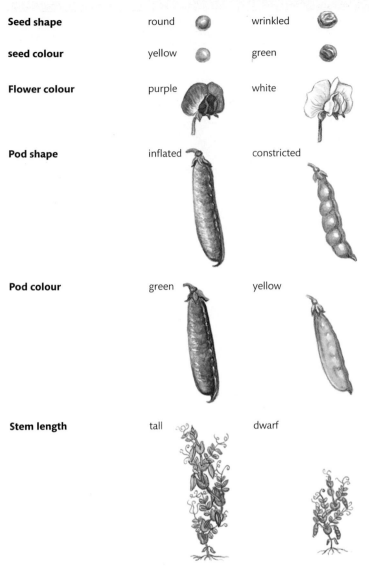

Seed shape	round		wrinkled
seed colour	yellow		green
Flower colour	purple		white
Pod shape	inflated		constricted
Pod colour	green		yellow
Stem length	tall		dwarf

Figure 12.2 *Some features of pea plants used by Mendel in his breeding experiments.*

Mendel decided to investigate, systematically, the results of cross-breeding plants that had contrasting features. These were the 'parent plants', referred to as '**P**' in genetic diagrams. He transferred pollen from one experimental plant to another. He also made sure that the plants could not be self-fertilised.

He collected all the seeds formed, grew them and noted the features that each plant developed. These plants were the first generation of offspring, or the '**F₁**' generation. He did not cross-pollinate these plants, but allowed them to self-fertilise. Again, he collected the seeds, grew them and noted the features that each plant developed. These plants formed the second generation of offspring or the '**F₂**' generation. When Mendel used pure-breeding tall and pure-breeding dwarf plants as his parents, he obtained the results shown in Figure 12.3.

In his breeding experiments, Mendel initially used only plants that had 'bred true' for several generations. For example, any tall pea plants he used had come from generations of pea plants that had all been tall.

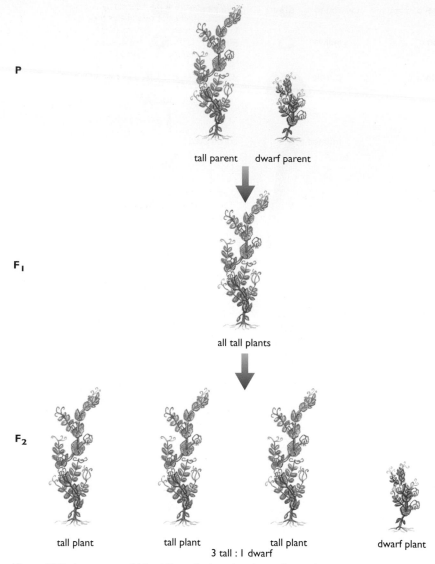

P

tall parent   dwarf parent

F_1

all tall plants

F_2

tall plant   tall plant   tall plant   dwarf plant

3 tall : 1 dwarf

Figure 12.3 *A summary of Mendel's results from breeding tall pea plants with dwarf pea plants.*

Mendel obtained very similar results when he carried out breeding experiments using plants with different pairs of contrasting characters (Figure 12.4). He noticed the following two things in particular.

- All the plants of the F_1 generation were always of just one type. This type was not a blend of the two parental features, but one or the other. Every time he repeated the experiment with the same feature, it was always the same type that appeared in the F_1 generation. For example, when tall and dwarf parents were cross-bred, the F_1 plants were always all tall.

- There was always a 3:1 ratio of types in the F_2 generation. Three-quarters of the plants in the F_2 generation were of the type that appeared in the F_1 generation. One-quarter showed the other parental feature. For example, when tall and dwarf parents were cross-bred, three-quarters of the F_2 plants were always tall and one-quarter were dwarf.

Mendel was able to use these patterns in his results to work out how features were inherited, without any knowledge of genes and chromosomes.

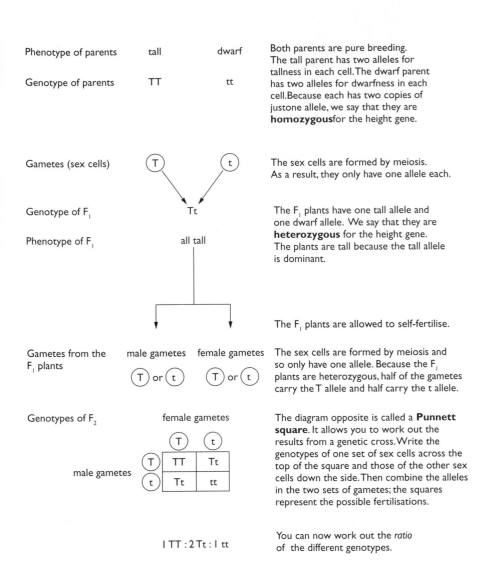

Phenotype of parents	tall	dwarf	Both parents are pure breeding. The tall parent has two alleles for tallness in each cell. The dwarf parent has two alleles for dwarfness in each cell. Because each has two copies of just one allele, we say that they are **homozygous** for the height gene.
Genotype of parents	TT	tt	

The sex cells are formed by meiosis. As a result, they only have one allele each.

Genotype of F₁ — Tt

The F₁ plants have one tall allele and one dwarf allele. We say that they are **heterozygous** for the height gene. The plants are tall because the tall allele is dominant.

Phenotype of F₁ — all tall

The F₁ plants are allowed to self-fertilise.

Gametes from the F₁ plants — male gametes (T) or (t)   female gametes (T) or (t)

The sex cells are formed by meiosis and so only have one allele. Because the F₁ plants are heterozygous, half of the gametes carry the T allele and half carry the t allele.

Genotypes of F₂

	female gametes	
	T	t
T	TT	Tt
t	Tt	tt

male gametes

The diagram opposite is called a **Punnett square**. It allows you to work out the results from a genetic cross. Write the genotypes of one set of sex cells across the top of the square and those of the other sex cells down the side. Then combine the alleles in the two sets of gametes; the squares represent the possible fertilisations.

1 TT : 2 Tt : 1 tt

You can now work out the *ratio* of the different genotypes.

Phenotypes of F₂ — 3 tall : 1 dwarf

Figure 12.4 *Results of crosses using true-breeding tall and dwarf pea plants.*

Whenever you have to work out a genetic cross, you should choose suitable symbols to represent the dominant and recessive alleles, give a key to explain which symbol is which and write out the cross exactly like the one shown.

Genotype describes the alleles each cell has for a certain feature, e.g. TT.
Phenotype is the feature that results from the genotype (e.g. a tall plant).

Explaining Mendel's results

We can now explain Mendel's results using the ideas of chromosomes, genes, mitosis and meiosis (Chapters 10 and 11).

- Each feature is controlled by a gene, which is found on a chromosome.

- There are two copies of each chromosome and each gene in all body cells, except the sex cells.

- The sex cells have only one copy of each chromosome and each gene.

- There are two alleles (forms) of each gene.

- One allele is **dominant** over the other allele, which is **recessive**.

- When two different alleles (one dominant and one recessive) are in the same cell, only the dominant allele is expressed (is allowed to 'work').

- An individual can have two dominant alleles, two recessive alleles or a dominant allele and a recessive allele in each cell.

The genotype of an organism is represented by two letters, each letter representing one allele of the gene that controls the feature. Normally, we use the initial letter of the dominant feature to represent the gene. Writing it as a capital letter indicates the dominant allele, the lower case letter represents the recessive allele. For the feature, height in pea plants, plants can be either tall or dwarf. Using the initial letter of the word 'tall', the alleles are shown as **T** and **t**. **TT** means that a plant has two alleles for tallness. A plant with two alleles for dwarfness would be represented by **tt**. Although we use pea plants to illustrate Mendel's ideas, the same principles apply to other organisms also.

We can use the cross between tall and dwarf pea plants as an example (Figure 12.4). In pea plants, there are tall and dwarf alleles of the gene for height. We will use the symbol **T** for the tall allele and **t** for the dwarf allele. The term **genotype** describes the alleles each cell has for a certain feature (e.g. TT). The **phenotype** is the feature that results from the genotype (e.g. a tall plant).

Ways of presenting genetic information

Writing out a genetic cross is a useful way of showing how genes are passed through one or two generations, starting with just two parents. To show a proper family history of a genetic condition requires more than this. We use a diagram called a **pedigree**. Polydactyly is an inherited condition in which a person develops an extra digit (finger or toe) on the hands and feet. It is determined by a dominant allele. The recessive allele causes the normal number of digits to develop.

If we use the symbol D for the polydactyly allele and d for the normal-number allele, the possible genotypes and phenotypes are:

- DD – person has polydactyly (has two dominant polydactyly alleles)

- Dd – person has polydactyly (has a dominant polydactyly allele and a recessive normal allele)

- dd – person has the normal number of digits (has two recessive, normal-number alleles).

We don't use **P** and **p** to represent the alleles as you would expect, because P and p look very similar and could easily be confused. The pedigree for polydactyly is shown in Figure 12.5.

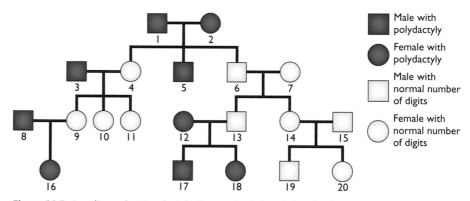

Figure 12.5 *A pedigree showing the inheritance of polydactyly in a family.*

We can extract a lot of information in a pedigree. In this case:

- there are four generations shown (individuals are arranged in four horizontal lines)

- individuals 4, 5 and 6 are children of individuals 1 and 2 (a family line connects each one directly to 1 and 2)

- individual 4 is the first-born child of 1 and 2 (the first-born child is shown to the left, then second born to the right of this, then third born and so on)

- individuals 3 and 7 are not children of 1 and 2 (no family line connects them directly to 1 and 2)

- 3 and 4 are father and mother of the same children – as are 1 and 2, 6 and 7, 8 and 9, 12 and 13, 14 and 15 (a horizontal line joins them).

It is usually possible to work out which allele is dominant from pedigrees. Look for a situation where two parents show the same feature and at least one child shows the contrasting feature. In this pedigree, 1 and 2 both have polydactyly, but children 4 and 6 do not. We can explain this in only one way:

- The normal alleles in 4 and 6 can only have come from their parents – 1 and 2, so 1 and 2 have normal alleles.

- 1 and 2 show polydactyly, so they *must* have polydactyly alleles as well.

- If they have both polydactyly alleles *and* normal alleles but show polydactyly, the polydactyly allele must be the dominant allele.

Now that we know which allele is dominant, we can work out most of the genotypes in the pedigree. All the people with the normal number of digits *must* have the genotype dd (if they had even one D allele, they would show polydactyly). All the people with polydactyly must have *at least one* polydactyly allele (they must be either DD or Dd).

From here, we can begin to work out the genotypes of the people with polydactyly. To do this we need to bear in mind that people with the normal number of digits must inherit one 'normal-number' allele from each parent, and also that people with the normal number of digits will pass on one 'normal-number' allele to each of their children.

From this we can say that any person with polydactyly who has children with the normal number of digits must be heterozygous (the child must have inherited one of their two 'normal-number' alleles from this parent), and also that any person with polydactyly who has one parent with the normal number of digits must also be heterozygous (the normal parent can only have passed on a 'normal-number' allele). Individuals 1, 2, 3, 16, 17 and 18 fall into one or both of these categories and must be heterozygous.

We can now add this genetic information to the pedigree. This is shown in Figure 12.6.

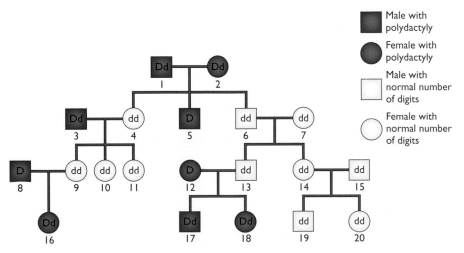

Figure 12.6 *A pedigree showing the inheritance of polydactyly in a family, with details of genotypes added.*

We are still left uncertain about individuals 5, 8 and 12. They could be homozygous or heterozygous. For example, individuals 1 and 2 are both heterozygous. Figure 12.7 shows the possible outcomes from a genetic cross between them. Individual 5 could be any of the outcomes indicated by the shading. It is impossible to distinguish between DD and Dd.

Genotypes of parents: Dd Dd

Gametes: (D) and (d) (D) and (d)

Genotypes of children

	female gametes (D)	female gametes (d)
male gametes (D)	DD	Dd
male gametes (d)	Dd	dd

Figure 12.7 *Possible outcomes from a genetic cross between two parents, both heterozygous for polydactyly.*

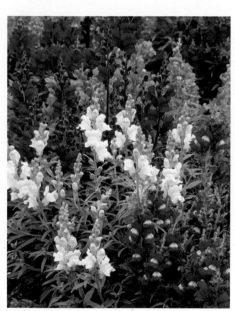

Figure 12.8 *Flower colours in snapdragons are caused by a gene showing codominance.*

Codominance

So far, all the examples of genetic crosses that we have seen involve **complete** dominance, where one **dominant** allele completely masks the effect of a second, or **recessive** allele. However, there are many genes with alleles that *both* contribute to the phenotype. If two alleles are expressed in the same phenotype, they are called **codominant**. For example, snapdragon plants have red, white or pink flowers (Figure 12.8).

If a plant with red flowers is crossed with one that has white flowers, all the plants resulting from the cross will have pink flowers. The appearance of a third phenotype shows that there is codominance. We can represent the alleles for flower colour with symbols:

R = allele for red flower

W = allele for white flower.

Figure 12.9 shows the cross between the parent plants. Note that the alleles for red and white flowers are given *different* letters, since one is not dominant over the other.

Genotypes of parent plants: RR WW

Gametes: all (R) all (W)

Genotypes of offspring all RW

	(R)	(R)
(W)	RW	RW
(W)	RW	RW

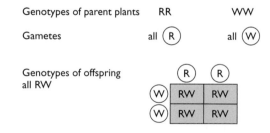

Figure 12.9 *Crossing red-flowered snapdragons with white-flowered plants produces a third phenotype, pink.*

When pink-flowered plants are crossed together, all three phenotypes reappear, in the ratio 1 red : 2 pink : 1 white (Figure 12.10).

Genotypes of parent plants   RW   RW

Gametes

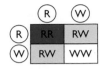

Genotypes of offspring
1RR : 2RW : 1WW

	R	W
R	RR	RW
W	RW	WW

Figure 12.10 *Crossing pink-flowered snapdragons.*

In fact, *most* genes do not show complete dominance. Genes can show a range of dominance, from complete dominance as in tall and dwarf pea plants through to equal dominance as in the snapdragon flowers, where the new phenotype is halfway between the other two.

ABO blood groups and codominance

The inheritance of human ABO blood groups also shows codominance. But the pattern of inheritance in blood groups is more complex than for flower colour in snapdragons, as three different alleles are involved. When there are more than two alleles of one gene, it is known as the inheritance of **multiple alleles**.

The blood group of a person is the result of the presence or absence of two antigens, the **A** antigen and the **B** antigen, on the surface of the red blood cells. There are three alleles involved in the inheritance of these antigens.

- I^A – determines the production of the **A** antigen

- I^B – determines the production of the **B** antigen

- I^o – determines that neither antigen is produced.

The alleles I^A and I^B are codominant, but I^o is recessive to both. Any one person can inherit only two alleles. The possible genotypes and phenotypes are shown in Table 12.1.

Genotype	Antigen produced	Blood group
$I^A I^A$, $I^A I^o$	A	A
$I^B I^B$, $I^B I^o$	B	B
$I^A I^B$	A and B	AB
$I^o I^o$	neither	O

Table 12.1: *Genotypes in blood groups.*

Parents who are heterozygous for blood group A and blood group B could produce four children, each with a different blood group (Figure 12.11).

Phenotypes   Blood   Blood
of parents   group A   group B

Genotypes $I^A I^o$ × $I^B I^o$
of parents

Gametes I^A and I^o I^B and I^o

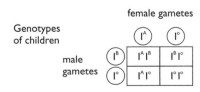

Genotypes
of children

	I^A	I^o
I^B	$I^A I^B$	$I^B I^o$
I^o	$I^A I^o$	$I^o I^o$

Phenotypes of children   IAB : IA : IB : IO

Figure 12.11 *Possible outcome from two parents heterozygous for blood groups A and B.*

You can interpret pedigrees of blood groups in the same way as other pedigrees. For example, in Figure 12.12, what are the blood groups of individuals 5 and 8?

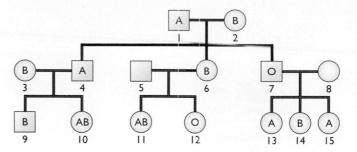

Figure 12.12 *Pedigree of human blood groups.*

Individual 12 has two Iº alleles (as she is blood group O) and so must inherit one of them from individual 5. Individual 11 must inherit her Iᴬ allele from individual 5, as her other parent is blood group B. Individual 5 therefore has the genotype IᴬIº and so must be blood group A.

Individuals 7 and 8 produce children with blood group A and blood group B. Individual 7 is blood group O and so both the Iᴬ and Iᴮ alleles must come from individual 8. Individual 8 is blood group AB.

Sex determination

Our sex – whether we are male or female – is not under the control of a single gene. It is determined by the X and Y chromosomes – the sex chromosomes. As well as the 44 non-sex chromosomes, there are two X chromosomes in all cells of females (except the egg cells) and one X and one Y chromosome in all cells of males (except the sperm). Our sex is effectively determined by the presence or absence of the Y chromosome. The full chromosome complement of male and female is shown in Figure 10.9 on page 143.

The inheritance of sex follows the pattern shown in Figure 12.13. In any one family, however, this ratio may well not be met. Predicted genetic ratios are usually only met when large numbers are involved. The overall ratio of male and female births in all countries is 1 : 1.

Sex-linked genes

The sex chromosomes don't just determine sex, they also carry genes for other characteristics. These are called **sex-linked genes**. The Y chromosome is smaller than the X, so it contains fewer genes. This means that for some genes, a male will only have one allele of a pair present on his X chromosome.

An example of this is the gene that causes a blood disorder called **haemophilia**. When a healthy person's skin is cut, a clot forms. This prevents loss of blood and entry of bacteria (see Chapter 4, page 63). Clotting is a complex process, involving many chemicals in the blood. The commonest type of haemophilia is caused by a gene mutation that affects the production of one of these chemicals – a protein in the blood plasma. Without this protein, the blood of a person with haemophilia does not clot. They may need blood transfusions after minor injuries, and injections of the missing clotting factor.

Because the Y chromosome, when present, causes a zygote to develop into a male, some people cannot resist describing it as 'dominant'. This is incorrect: dominant and recessive are terms that are only applied to individual alleles.

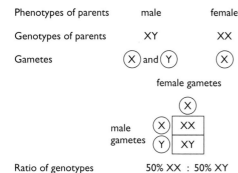

Phenotypes of parents	male	female
Genotypes of parents	XY	XX
Gametes	(X) and (Y)	(X)

female gametes

		(X)
male gametes	(X)	XX
	(Y)	XY

Ratio of genotypes	50% XX : 50% XY
Ratio of phenotypes	50% female : 50% male

Figure 12.13 *Determination of sex in humans.*

The allele for haemophilia is recessive, and is given the symbol h. The allele for normal blood clotting is dominant (H). Since the gene is found only on the X chromosome, a man needs to inherit only one allele of the gene to have the disease. The genotype for this is shown as X^hY – notice there is no allele for this gene on the Y chromosome. A woman, with two X chromosomes, would need to inherit two copies of the faulty allele, which is shown as X^hX^h. If a woman has only one copy of the haemophilia allele (X^HX^h), she will not have the disease, because of the presence of the dominant H allele on one of her X chromosomes. However, she can pass the haemophilia allele to her children, so she is called a **carrier**.

Boys normally inherit the recessive allele from a carrier mother. This means it is possible for two healthy parents to have a son with haemophilia (Figure 12.14).

There are several types of haemophilia. They all have a genetic cause and are due to a lack of blood-clotting factors. Two forms are due to a sex-linked gene on the X chromosome.

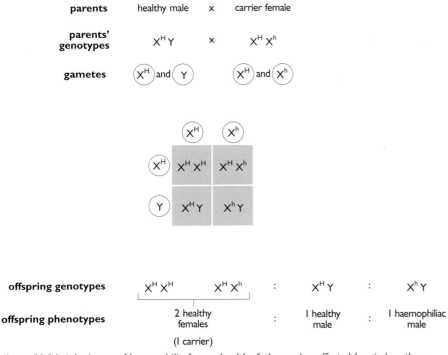

Figure 12.14 *Inheritance of haemophilia from a healthy father and unaffected (carrier) mother.*

For a girl to have haemophilia, she would have to be the daughter of a haemophiliac father and a carrier mother (Figure 12.15). This is possible, but much less likely. This means that haemophilia is much more common in boys than girls. The most common form of haemophilia affects about one in 5 000–10 000 males.

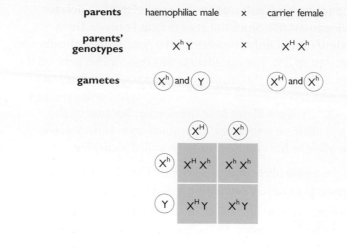

	X^H	X^h
X^h	$X^H X^h$	$X^h X^h$
Y	$X^H Y$	$X^h Y$

offspring genotypes	$X^H X^h$	:	$X^h X^h$	:	$X^H Y$	:	$X^h Y$
offspring phenotypes	I healthy female	:	I haemophiliac female	:	I healthy male	:	I haemophiliac male
	(carrier)						

Figure 12.15 *Inheritance of haemophilia from a haemophiliac father and unaffected (carrier) mother.*

There are several other sex-linked genes like this. **Red-green colour blindness** is another example. It is inherited in the same way as haemophilia, and is more common in boys than girls. Of course, it is a much less serious condition than haemophilia.

Other genetic diseases

Faulty genes are responsible for many other conditions that we call genetic diseases. Two of these are **cystic fibrosis** and **sickle cell anaemia**.

Cystic fibrosis

This condition is determined by a recessive allele of the gene that controls the production of mucus by cells in glands throughout the body. The gene has a high mutation rate, and if mutations occur in the sex cells (or in the cells that form the sex cells) the mutant allele could be inherited.

The normal, dominant allele results in normal mucus being secreted. The mutated, recessive allele causes the production and secretion of viscous (very thick) mucus. This has several adverse effects:

- It blocks the pancreatic duct so that pancreatic enzymes cannot reach the small intestine. This affects the digestion of carbohydrates, lipids and proteins (see Chapter 3).

- It cannot be easily moved out of the lungs by the cilia as can normal mucus. Gas exchange suffers as a result.

Because of these effects, people suffering from cystic fibrosis often die young. However, treatment is now much better and can extend their lifespan. Gene therapy may offer a cure in the future.

To be affected, a person must inherit two recessive alleles (one from each parent), so each parent must carry at least one recessive cystic fibrosis allele. The disease is inherited most commonly when both parents are heterozygous for the condition (Figure 12.16). There is a one in four (25%) chance of a child from such a relationship developing cystic fibrosis.

Sickle cell anaemia

This disease is also caused by a mutant allele of a gene. With sickle cell anaemia, the gene codes for the production of haemoglobin in red blood cells. People with the disease produce abnormal haemoglobin. When the oxygen concentration of their blood is low, the abnormal haemoglobin causes the red blood cells to change from their normal biconcave disc shape into a sickle shape that gives the disease its name (Figure 12.17). The disease is common in areas of the world where malaria is found (see below), particularly in central and southern Africa, and equatorial regions of South America. In Africa south of the Sahara Desert, about one-third of all black people carry the sickle cell allele. The gene also affects people of African descent living in other parts of the world. For example, in the United States about one in 500 babies of black people of African descent are born with the disease.

There are two main consequences of the red blood cells becoming sickle shaped.

- The sickle cells stick together, forming blockages in the capillaries and causing severe pain, especially in the joints. This is known as a sickle cell crisis. Several organs may be affected by the reduced blood supply. If the blood supply to the brain is affected, a stroke may result.

- The sickle shape reduces the amount of oxygen the blood cells can carry. Sickle cells often burst, and are destroyed by the spleen at a higher rate than normal cells. The reduction in the number of red blood cells causes the anaemia that gives the disease the other part of its name.

Sickle cell anaemia is a dangerous condition, and people with the disease have a reduced life expectancy, although its symptoms and effects can be controlled by medical treatment. Following a sickle cell crisis, a blood transfusion can be given to replace the abnormal cells with normal ones. Bone marrow is the source of red blood cells, and bone marrow transplants have also been used to treat the disease. If bone marrow from a non-sufferer is successfully transplanted into a person with the disease, it will produce normal red blood cells. These are expensive treatments and only available in specialist centres, so are not available to poor people in the many countries where the disease is common.

The sickle cell gene is codominant with the gene for normal haemoglobin (see page 164). The symbol for the allele for normal haemoglobin is Hb^A, and the symbol for the allele for sickle cell haemoglobin is Hb^S. So there are three possible genotypes: Hb^{A}Hb^A, Hb^{A}Hb^S and Hb^{S}Hb^S. Table 12.2 compares the phenotypes and effects of these three genotypes.

N = dominant normal allele resulting in production of normal mucus

n = recessive cystic fibrosis allele resulting in the production of viscous mucus

Heterozygous non-sufferer = Nn

Genotypes of parents   Nn   ×   Nn

Gametes   (N) and (n)   (N) and (N)

female gametes

		N	n
male gametes	N	NN	Nn
	n	Nn	nn

this child would develop cystic fibrosis

Figure 12.16 *Inheritance of cystic fibrosis.*

A sickle is a crescent-shaped agricultural tool used to cut down vegetation.

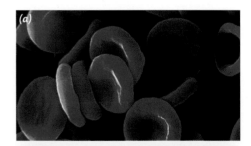
(a)

(b)

Figure 12.17 *(a) Normal red blood cells; (b) distorted red blood cells from a person suffering from sickle cell anaemia.*

Anaemia is any condition in which the concentration of haemoglobin in the blood falls below normal.

Genotype	Hb^AHb^A	Hb^AHb^S	Hb^SHb^S
Phenotype	normal red blood cells	mostly normal red blood cells, some sickle-shaped	many sickle-shaped red blood cells
Effects	healthy	normally no symptoms except slight anaemia	sickle cell anaemia with severe sickle cell crises

Table 12.2: *Genotypes and phenotypes involving the allele for sickle cell haemoglobin.*

Notice that the heterozygote Hb^AHb^S is intermediate in phenotype between the other two genotypes. One allele does not show complete dominance over the other – they are codominant. A man who is heterozygous for the gene (a carrier) will generally be healthy. However, if he has a child with a woman who is also heterozygous, there is a one in four chance that the child will have the disease (Figure 12.18).

It is important to note that the gene for sickle cell anaemia is not on the sex chromosome, in other words it is not sex linked. The probability of a boy or a girl from the same parents inheriting the disease will be the same.

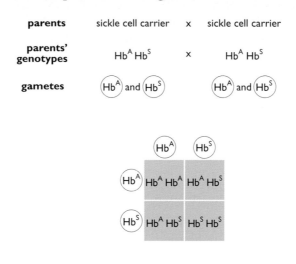

Figure 12.18 *A cross between two people, both heterozygous carriers of the sickle cell allele.*

Sickle cell anaemia and natural selection

Sickle cell anaemia lowers a person's life expectancy, and many children with the disease die before they reach maturity, so you might expect this to be a strong disadvantage that would act to remove the sickle cell gene from the population through the process of natural selection (see Chapter 10, page 145).

However, heterozygous carriers of the sickle cell allele usually show no symptoms of the disease at all, despite about 40% of the haemoglobin in their red blood cells being abnormal. In fact, they can actually benefit from their condition. This is because being a carrier gives them increased resistance to malaria. Malaria is a disease that kills millions of people around the world every year (see Chapter 13, page 186), and it is no coincidence that the malaria zones of the world overlap with the areas where sickle cell anaemia is common (Figure 12.19).

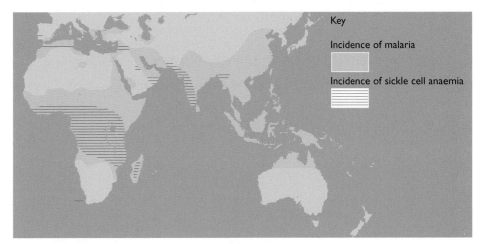

Figure 12.19 *A map showing areas of the world where sickle cell anaemia and malaria are common.*

How does being Hb^AHb^S protect against malaria? The parasite that causes malaria spends part of its life cycle inside red blood cells. The red blood cells of carriers look normal, but because of the 40% abnormal haemoglobin they contain, they are slightly more fragile than normal cells. When the parasite enters the fragile red blood cells of a carrier, the cells often burst before the parasite has had time to develop, and the parasite dies. The life cycle is broken. This means that sickle cell carriers are at an advantage in malaria zones – they are more likely to survive and reproduce, passing on the Hb^S allele to their children. In terms of Darwin's natural selection, they have a selective advantage over individuals who do not carry the allele. Over many generations, natural selection has acted to maintain the sickle cell gene in the population, through the advantage shown by the heterozygote.

End of Chapter Checklist

You should now be able to:

✓ recall that genes exist in alternative forms called alleles, which can be dominant or recessive

✓ recall the meaning of the terms dominant, recessive, homozygous, heterozygous, phenotype and genotype

✓ understand patterns of monohybrid inheritance using a genetic diagram

✓ understand how to interpret family pedigrees

✓ understand the inheritance of the ABO blood groups, and recall the terms multiple alleles and codominance

✓ recall that the sex of a person is controlled by one pair of chromosomes, XX in a female and XY in a male

✓ explain how to determine the sex of offspring at fertilisation, using a genetic diagram

✓ understand how to predict probabilities of outcomes from monohybrid crosses, including cystic fibrosis and sickle cell anaemia

✓ understand that mutations can increase in a population by natural selection

Questions

1 Predict the *ratios* of the phenotypes of offspring from the following crosses between tall/dwarf pea plants.

a) TT · TT, **b)** TT · Tt, **c)** TT · tt, **d)** Tt · Tt, **e)** Tt · tt, **f)** tt · tt.

2 Cystic fibrosis is an inherited condition. The diagram shows the incidence of cystic fibrosis in a family over four generations.

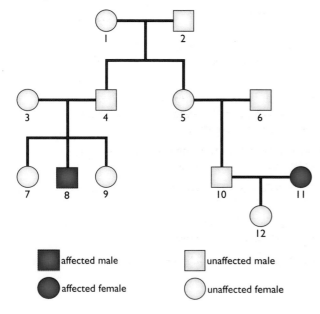

affected male

affected female

unaffected male

unaffected female

a) What evidence in the pedigree suggests that cystic fibrosis is determined by a recessive allele?

b) What are the genotypes of individuals 3, 4 and 11? Explain your answers.

c) Draw genetic diagrams to work out the probability that the next child born to individuals 10 and 11 will i) be male, ii) suffer from cystic fibrosis.

3 The diagram shows the inheritance of phenylthiocarbamide (PTC) tasting in a family. Although PTC has a very bitter taste, some people cannot taste it.

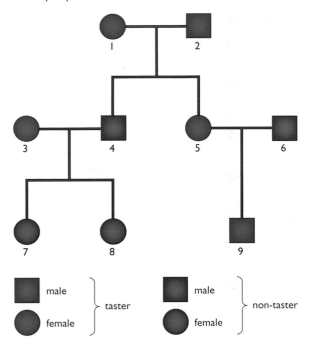

male — taster
female — taster
male — non-taster
female — non-taster

a) What evidence in the diagram suggests that the allele for PTC tasting is dominant?

b) Using T to represent the tasting allele and t to represent the non-tasting allele, give the genotypes of individuals 3 and 7. Explain how you arrived at your answers.

c) Why can we not be sure of the genotype of individual 5?

d) If individuals 3 and 4 had another child, what is the chance that the child would be able to taste PTC? Construct a genetic diagram to show how you arrived at your answer.

4 Sickle cell anaemia is determined by a single mutant allele. Sufferers are homozygous for this allele. Heterozygotes show virtually no signs of the condition, but have an increased resistance to malaria.

a) What is a sickle cell crisis?

b) Why does a bone marrow transplant often cure sickle cell anaemia?

c) A sickle cell sufferer survived to have children and married a person heterozygous for the condition. What proportion of their children would develop sickle cell anaemia? Use a genetic diagram to explain your answer.

d) Why do people heterozygous for the condition have an increased resistance to malaria?

e) In the United States, the occurrence of sickle cell anaemia among black people of African ancestry is lower than in Africa, and is falling with each generation. There is no malaria in the United States. Explain this in terms of natural selection.

5 The inheritance of ABO blood groups is controlled by three alleles of the same gene. The alleles are given the symbols I^A, I^B and I^O. Alleles I^A and I^B are codominant. Both alleles I^A and I^B are dominant to allele I^O.

a) Explain what is meant by the terms:

  i) allele

  ii) codominant.

b) Complete the table to show all the possible genotypes for each blood group. One has been done for you.

Blood group	Genotype
A	$I^A I^A$
B	
AB	
O	

c) A woman with blood group A married a man with blood group B. They had one child with blood group A, and another child with blood group O. Construct a genetic diagram to show the results of this cross. Indicate the phenotypes of the children on your diagram.

6 A couple you know have four children, all girls. They insist that their next child will be a boy. Do you agree with them? What is the probability that their next child will be a boy? Draw a diagram to explain this.

7 One of the genes for colour vision is found on the X chromosome, but is not present on the Y chromosome. The dominant allele of this gene produces normal colour vision, while the recessive allele causes red-green colour blindness.

a) Using the symbol X^A for a dominant allele on the X chromosome, X^a for a recessive allele on the X chromosome, and Y for the Y chromosome, write down the genotypes of:

  i) a colour-blind man

  ii) a colour-blind woman

  iii) a woman who is a carrier for the colour blindness gene.

b) A couple who both have normal colour vision have a child with red-green colour blindness. Draw a diagram to show how this came about. What sex is the child?

Chapter 13: Microorganisms and Disease

You have seen that some diseases are caused by an unhealthy lifestyle, such as smoking, drinking alcohol or eating a poor diet. Other diseases are genetic, caused by faulty genes. These are both examples of non-infectious diseases. Some diseases are infectious – they are caused by microorganisms. This chapter looks at the types of microorganism that can cause disease, how they enter the body, and how the body responds to the infection.

What is a disease?

It is not easy to define the word 'disease'. Disease doesn't just mean the absence of health – if you are less fit than you could be, or can't sleep because you are thinking about an approaching exam, you might be considered unhealthy, but this doesn't mean you have a disease. One definition of disease is 'a condition with a specific cause in which part or all of the body functions abnormally and less efficiently'. The cause can be unhealthy activities such as smoking or drinking alcohol; it can be genetic, such as the mutated genes responsible for haemophilia or sickle cell anaemia; or, most commonly, the cause is a **microorganism**. Microorganisms are responsible for causing **infectious** diseases – those that can be transmitted from one person to another.

Microorganisms that cause diseases are known as **pathogens** or pathogenic organisms. There are many different types, including bacteria, viruses, fungi and protozoa. Table 13.1 and Figure 13.1 give details of some of these.

Type of microorganism	How the microorganism causes disease	Examples of diseases caused
bacteria	Bacteria release toxins (poisons) as they multiply. The toxins affect cells in the region of the infection, and sometimes in other parts of the body as well.	typhoid, tuberculosis (TB), gonorrhoea, cholera, pneumonia
viruses	Viruses enter a living cell and disrupt the metabolic systems of that cell. The genetic material of the virus takes over the cell and instructs it to produce more virus.	influenza ('flu), poliomyelitis (polio), human immunodeficiency virus (HIV), measles, rubella, common cold
fungi	When fungi grow in or on the body, their fine threads (hyphae) secrete digestive enzymes onto the tissues, breaking them down. Growth of hyphae also physically damages body tissues. Some fungi secrete toxins. Others cause an allergic reaction.	thrush, athlete's foot, ringworm (a skin disease), 'farmer's lung'
protozoa	(there is no set pattern as to how protozoa cause disease)	malaria, trypanosomiasis (sleeping sickness)

Table 13.1: *The main types of pathogenic microorganism.*

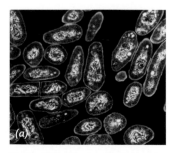 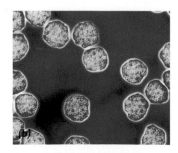 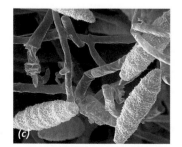 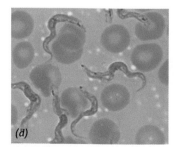

Figure 13.1 *Some pathogenic microorganisms: (a) the bacterium that causes tuberculosis; (b) the measles virus; (c) the mould fungus responsible for ringworm; (d) the protozoan* Trypanosoma gambiense *– the cause of sleeping sickness.*

Types of microorganism

Bacteria

Bacteria are very small, single-celled organisms. To give you some idea of their size, a typical animal cell might be 10–50 μm in diameter (1 μm, or one micrometre, is a millionth of a metre, or a thousandth of a millimetre). Compared with this, a typical bacterium is only 1–5 μm in length (Figure 13.2), and its volume is thousands of times less than the larger cell.

There are three basic shapes of bacteria: spheres, rods and spirals, but they all have a similar internal structure (Figure 13.3).

(a) **Some different bacterial shapes**

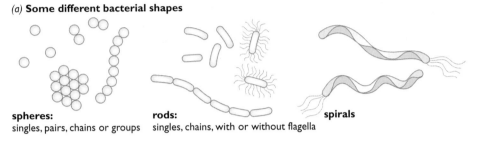

spheres:
singles, pairs, chains or groups

rods:
singles, chains, with or without flagella

spirals

(b) **Internal structure of a bacterium**

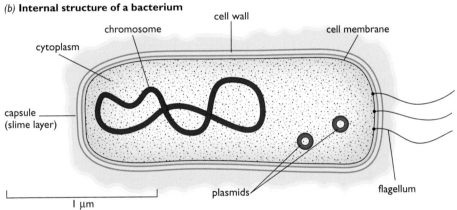

Figure 13.3 *Shapes and structure of bacteria.*

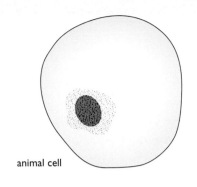

animal cell

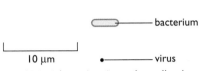

bacterium

10 μm

virus

Figure 13.2 *A bacterium is much smaller than an animal cell. The relative size of a virus is also shown.*

All bacteria are surrounded by a cell wall, which protects the bacterium and keeps the shape of the cell. Unlike in plants, bacterial cell walls are not made of cellulose, but are composed of complex chemicals made of polysaccharides and proteins. Some species have another layer outside this wall, called a **capsule** or **slime layer**. Both give the bacterium extra protection. Underneath the cell wall is the cell membrane, as in other cells. The middle of the cell is made of cytoplasm. One major difference between a bacterial cell and the more complex cells of animals and plants is that the bacterium has no nucleus. Instead, its genetic material (DNA) is in a **single chromosome**, loose in the cytoplasm, forming a circular loop.

Some bacteria can swim, and are propelled through water by corkscrew-like movements of structures called **flagella** (a single one of these is called a flagellum). However, many bacteria do not have flagella and cannot move by themselves. Other structures present in the cytoplasm include the **plasmids**. These are small, circular rings of DNA, carrying some of the bacterium's genes. Not all bacteria contain plasmids, although about three-quarters of all known species do.

Most species of bacteria are harmless to humans. Some free-living species contain a form of chlorophyll in their cytoplasm, and can carry out photosynthesis. Most bacteria, along with fungi, are important **decomposers** (see Chapter 14), recycling dead organisms and waste products in the soil and in sewage treatment. Some bacteria are even used by humans to make food, such as *Lactobacillus bulgaricus*, a rod-shaped species used in the production of yoghurt (Figure 13.4). Only relatively few species are pathogens (Figure 13.5).

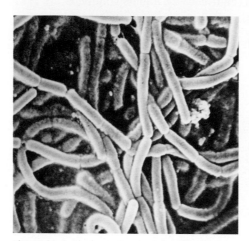

Figure 13.4 *The bacterium* Lactobacillus bulgaricus, *used in the production of yoghurt.*

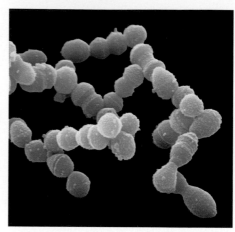

Figure 13.5 *Rounded cells of the bacterium* Pneumococcus, *one cause of pneumonia.*

Despite the relatively simple structure of the bacterial cell, it is still a living cell that carries out the normal processes of life – respiration, feeding, excretion, growth and reproduction. Compare this with the next group, the much simpler viruses.

Viruses

All viruses are parasites, and can only reproduce inside living cells. The cell in which the virus lives is called the **host**. Viruses are much smaller than bacterial cells: most are between 0.01 and 0.1 μm in diameter (see Figure 13.2). Very few viruses infect bacterial cells.

Notice that we say 'types' of virus, not 'species'. This is because viruses are not made of cells. A virus particle is very simple. It has no nucleus or cytoplasm, and is composed of a core of genetic material surrounded by a protein coat (Figure 13.6). The genetic material can be either DNA or RNA (see Chapter 10). In either case, the genetic material makes up just a few genes – all that is needed for the virus to reproduce inside its host cell.

Sometimes a membrane called an envelope may surround a virus particle, but the virus does not make this. Instead it is 'stolen' from the surface membrane of the host cell.

Viruses do not feed, respire, excrete, move, grow or respond to their surroundings. They do not carry out any of the normal characteristics of living things except reproduction, and they can only do that parasitically. This is why some scientists view viruses as being on the border between a living organism and a non-living chemical.

A host is an organism that is infected by another, smaller organism.

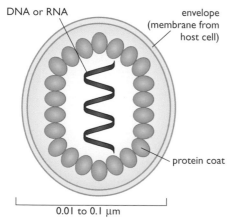

Figure 13.6 *The structure of a typical virus, such as the type causing influenza ('flu).*

A virus reproduces by entering the host cell and taking over the host's genetic machinery to make more virus particles. After many virus particles have been made, the host cell dies and the particles are released to infect more cells. Many human diseases are caused in this way, including colds, 'flu, measles, mumps, polio and rubella. Of course, the reproduction process does not go on forever. Usually the body's immune system destroys the virus and the person recovers. Sometimes, however, a virus cannot be destroyed by the immune system quickly enough, and it may cause permanent damage or death. With other infections, the virus may attack cells of the immune system itself. This is the case with HIV (human immunodeficiency virus), which eventually causes the disease AIDS (acquired immune deficiency syndrome). We will look at the HIV virus in more detail later in this chapter.

Fungi

Most fungi are free-living, and do not cause disease. They include mushrooms and toadstools, as well as moulds. These groups of fungi are multicellular. Another group of fungi are the yeasts, which are made of single cells. Fungi feed by secreting enzymes onto organic material, breaking it down, and absorbing the soluble products. Nearly all fungi feed on dead organic material. Along with bacteria, they are important decomposers in ecosystems (see Chapter 14).

A few species of moulds and yeasts are pathogens. For example, the disease thrush is caused a species of yeast that lives on the mucous membranes of the body, such as the mouth (see later in this chapter). 'Athlete's foot' is a skin disease caused by a fungus feeding on the skin. 'Farmer's lung' is a lung disease caused by inhaling spores of a mould.

Protozoa

Protozoa are single-celled organisms. They consist of animal-like cells, and contain the cell organelles you would expect to find in a typical animal cell – nucleus, cytoplasm, mitochondria, etc. Protozoa are mostly free-living, inhabiting almost any type of wet or moist habitat, such as ponds, rivers and the sea. One of the best known is an organism called *Amoeba*, which lives in ponds (Figure 13.7).

There a few diseases that are caused by pathogenic protozoa. For example, a pathogenic species of *Amoeba* causes an intestinal disease called dysentery. Although the number of protozoan diseases is relatively small, some are very serious indeed. They include sleeping sickness, caused by a protozoan called a trypanosome (see Figure 13.1d), and malaria, caused by a species called *Plasmodium* (see later in this chapter).

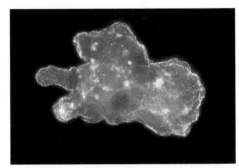

Figure 13.7 *Amoeba, a protozoan living in ponds.*

Some infectious diseases are caused by larger organisms, particularly worms, such as roundworms or tapeworms. One widespread disease caused by a species of worm is schistosomiasis (bilharzia) – see later in this chapter.

Transmission of diseases

Transmission of a disease is the means by which the pathogen is transferred from one host to another. Table 13.2 summarises the main methods of transmission of infectious diseases.

Method of transmission	How the transmission route works	Examples of diseases
droplet infection	Many of these are respiratory diseases (diseases that affect the airways of the lungs). The organisms are carried in tiny droplets through the air when an infected person coughs or sneezes. They are inhaled by other people.	common cold, influenza, tuberculosis, pneumonia
drinking contaminated water	The microorganisms transmitted in this way often infect regions of the gut. When a person drinks unclean water containing the organisms, they colonise a suitable area of the gut and reproduce. They are passed out with faeces and find their way back into the water.	cholera, typhoid fever, polio
eating contaminated food	Most food poisoning is bacterial, but some viruses are transmitted this way. The organisms initially infect a region of the gut.	typhoid fever, polio, salmonellosis, listeriosis, botulism
direct contact	Many skin infections, such as athlete's foot, are spread by direct contact with an infected person or contact with a surface carrying the organism.	athlete's foot, ringworm
sexual intercourse	Organisms infecting the sex organs can be passed from one sexual partner to another during intercourse. Some (such as the fungus thrush, which causes candidiasis) are transmitted by direct body contact. Others (such as the AIDS virus) are transmitted in semen or vaginal secretions. Some (such as syphilis) can be transmitted in saliva.	chlamydia, syphilis, AIDS, gonorrhoea
blood-to-blood contact	Many sexually transmitted diseases can also be transmitted in this way. Drug users sharing an infected needle can transmit AIDS.	AIDS, hepatitis B
animal vectors	Many diseases are transmitted by insect bites. Mosquitoes spread malaria and tsetse flies spread sleeping sickness. In both cases, the disease-causing organism is transmitted when the insect bites humans in order to suck blood. Also, flies can carry microorganisms from faeces onto food.	malaria, sleeping sickness, typhoid fever, salmonellosis

Table 13.2: *Methods of disease transmission.*

Figure 13.8 *A fly feeding on human food. The fly releases saliva onto the food as it feeds. The saliva may contain disease-causing organisms.*

Insects such as the housefly are responsible for transmitting many diseases. Houseflies are attracted to animal or human faeces, and transmit many bacteria and viruses on their body or in their saliva. They do not bite humans, but will feed on human food if it is left uncovered (Figure 13.8). As the fly feeds, it releases its saliva onto the food. The pathogenic microorganisms may then be transferred to humans when they eat the food. Many serious diseases, such as diphtheria, meningitis, typhoid, cholera and polio, are transmitted in this way.

The general course of a disease

After a person has been infected with a disease, there is an **incubation period**. This is the time between when a person is first infected with the pathogen and when they first show signs and symptoms of the disease. During the incubation period, the infected person may not feel sick, but could be infectious to others. Incubation periods vary greatly between diseases, from hours to months. Some diseases like leprosy can have an incubation period lasting years.

'Signs' and 'symptoms' of a disease have slightly different meanings.

- A **sign** of a disease is visible to other people. It can be seen, heard or measured. For example, a doctor might listen to a patient's chest with a stethoscope to hear signs of a chest infection, or measure their blood pressure to check for a heart problem.

- A **symptom** is not usually visible to other people. It is what the patient is experiencing as a result of the disease, such as pain, chills, dizziness or nausea. The symptoms are the first thing that a patient notices that makes them go to the doctor.

Distribution of diseases

Some diseases are only found in certain parts of the world. For example, malaria is common in tropical and subtropical countries (Figure 13.9). If a disease is always present in a population of a particular geographical area, it is **endemic**.

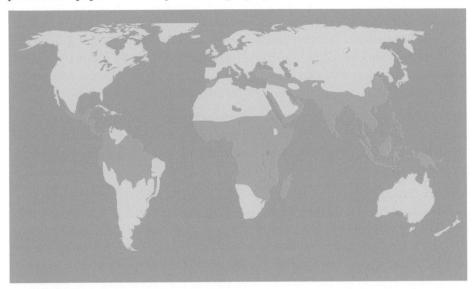

☐ areas in which malaria is endemic

Figure 13.9 *Regions of the world where malaria is endemic.*

This is not to be confused with an **epidemic** – which is a widespread outbreak of an infectious disease, with many people becoming infected at the same time, spreading over a wide area. For a disease to be classified as an epidemic, there must be an increase in the area it affects, not just the numbers infected. If the disease spreads across the world, we call it a **pandemic**. There have been 'flu pandemics in 1918 (Spanish 'flu), 1957 (Asian 'flu), 1968 (Hong Kong 'flu) and 2009 (swine 'flu).

The expansion in worldwide travel over the past 30 years has exposed many people to diseases that are not normally endemic in their own country. Potentially, this can increase the spread of diseases from an area where they are currently endemic to other new areas.

It has been estimated that over 50 million people died from the Spanish 'flu pandemic in 1918-19 – more than were killed in the First World War of 1914-18.

Important infectious diseases

In this section we will look at a range of infectious diseases. You will read about three diseases caused by viruses (influenza, poliomyelitis and AIDS); three that are bacterial (typhoid, tuberculosis and gonorrhoea); two due to fungal infections (thrush and athlete's foot); one disease caused by a protozoan (malaria); and one disease caused by a parasitic worm (schistosomiasis). Of course, these are only a few of the hundreds of infectious human diseases that exist. They are included as examples – the main thing for you to understand is how they are caused and how they are transmitted from person to person, as well as how they can be treated and prevented from spreading.

Influenza

Influenza ('flu) is caused by a virus (Figure 13.10). The virus primarily infects cells in the upper airways of the respiratory system. Sometimes it spreads down the trachea and into the lungs. 'Flu is transmitted by airborne droplets produced when an infected person coughs or sneezes.

Symptoms of infection include chills, fevers and muscular aches, as well as loss of appetite and fatigue. The initial symptoms are often followed by a cough, sore throat and runny nose. The fever and aches usually subside after about 5 days, but the other symptoms last longer and the person may feel weak for about 10 days. If the disease does affect the lungs, it causes an acute form of pneumonia that can be fatal within a few days, even in otherwise healthy people.

Influenza can be made worse if the person's breathing system is also infected by bacteria after the 'flu virus. This is called a secondary infection. Bacteria causing bronchitis and pneumonia often cause secondary infections in children and elderly people with 'flu.

There are three strains of the 'flu virus – the A, B and C strains. Type C produces a much milder form of 'flu than the other two strains. The A and B strains mutate periodically and produce new strains. This makes everyone open to infection, even if they have had 'flu before. The antibodies that were made to the last form of 'flu (see later in this chapter) will not be effective against the new strains. For the same reason, new vaccines are needed to combat each new strain of 'flu.

There is little that can be done to treat the illness, apart from trying to ease the painful symptoms. Antibiotics do not work against viruses, so the body must destroy the virus by itself. But antibiotics may be prescribed to combat secondary bacterial infections. The spread of the disease can be controlled to some extent by vaccination, but current 'flu vaccines are only about 60–70% effective. Vaccination is targeted towards groups of people who need it most, such as the elderly. People with 'flu should stay away from other people while they are infectious.

Poliomyelitis

Poliomyelitis (polio) is an infectious viral disease spread from person to person, often by flies carrying the virus from human faeces to food. Most people infected with polio show no symptoms, but in about 1% of people the virus enters the brain or spinal cord, where it destroys nerve cells, particularly motor neurones. The damage leads to muscle weakness and paralysis of the limbs (Figure 13.11). It can also paralyse the muscles of the breathing system resulting in death.

Polio seems to have existed as an endemic disease until the late nineteenth century – it was always present, but rarely developed into an epidemic. In the early twentieth century, major epidemics began to occur in Europe and the United States, and spread throughout the world, causing paralysis and death to hundreds of thousands of people, particularly young children. Eventually a vaccine against polio was developed in the 1950s. Nowadays, vaccination has reduced the incidence of polio from hundreds of thousands to just thousands of cases.

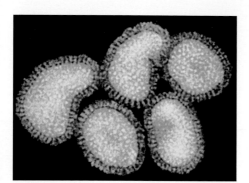

Figure 13.10 *The influenza virus.*

Mutations alter one or more genes in the virus. Some mutations cause the virus to produce different antigens on its surface.

Figure 13.11 *This Afghan boy has paralysed legs as a result of polio – his leg muscles have withered. Afghanistan is one of the few countries still trying to eradicate polio.*

The other main method of prevention is good hygiene, including modern sanitation so that flies do not come into contact with human sewage, as well as preventing them from coming into contact with human food. The World Health Organization hopes that some time in the near future enhanced vaccination programmes will result in complete eradication of the disease.

AIDS

AIDS (acquired immune deficiency syndrome) is currently one of the most significant worldwide killers. It is caused by a virus called the human immunodeficiency virus (**HIV**). Figure 13.12 shows the structure of this virus.

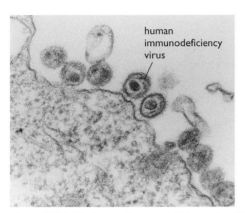
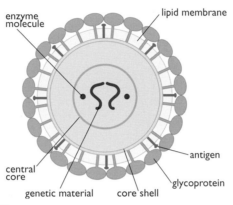

Figure 13.12 *The human immunodeficiency virus (HIV).*

HIV initially infects a type of white blood cell called a helper T-lymphocyte. These cells are necessary if other white blood cells (B-lymphocytes and other T-lymphocytes) are to become active and start fighting infections (see later in this chapter). The course of a typical HIV infection is described below.

1. The genetic material of HIV becomes incorporated in the DNA of the helper T-lymphocyte.

2. When the HIV DNA is activated, it 'instructs' the lymphocyte to make HIV proteins and genetic material.

3. Some of the HIV proteins and genetic material are assembled into new viruses.

4. Some of the HIV proteins end up as marker proteins (antigens) on the surface of the cell.

5. These HIV proteins are recognised as foreign proteins.

6. The lymphocyte is destroyed by the immune system.

7. The assembled virus particles escape to infect other lymphocytes.

8. This cycle repeats itself for as long as the body can replace the lymphocytes that have been destroyed.

9. Eventually the body will not be able to replace the lymphocytes at the same rate at which they are being destroyed.

10. The number of free viruses in the blood increases rapidly and HIV may infect other areas of the body, including the brain.

There was once a disease unique to humans called smallpox. It was a virus disease, producing a characteristic rash and blisters over the skin, and estimated to have caused up to 500 million deaths around the world during the twentieth century. A worldwide vaccination programme against smallpox was set in motion, and in December 1979 the disease was officially declared to have been eradicated (wiped out). No more cases have been found since then. So far, this is the only human disease that has been completely eradicated in this way.

Part of this immune response to infection by HIV involves the production of antibodies to destroy the virus. At this stage, the person is said to be **HIV positive** because their blood gives a positive result when tested for HIV antibodies.

The period during which the body replaces the lymphocytes as fast as they are destroyed is called the **latency period**. It can last for up to 20 years. The person shows no symptoms of AIDS during this period, but will be highly infective to others.

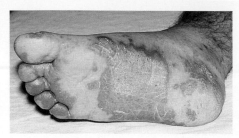

Figure 13.13 *The foot of a person suffering from Kaposi's sarcoma.*

11. The immune system is severely damaged, and other disease-causing microorganisms infect the body.

12. Death is often as a result of 'opportunistic' infection by TB and pneumonia, due to the reduced capacity of the immune system that would normally destroy the organisms causing these diseases. Death can also be caused by rare cancers such as Kaposi's sarcoma (Figure 13.13).

Figure 13.14 shows how the levels of virus particles, HIV antibodies and helper T-lymphocytes in the blood change during the course of an AIDS infection.

The real significance of AIDS is that the cells being destroyed are the cells needed to help the other lymphocytes destroy the infected cells.

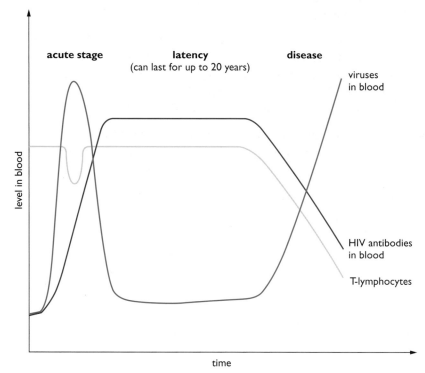

Figure 13.14 *The stages of an AIDS infection.*

AIDS is primarily transmitted by unprotected sexual intercourse (either homosexual or heterosexual), or by blood-to-blood contact (for example, when drug users share an infected needle, or if infected blood is given in a transfusion). AIDS can *not* be transmitted by kissing, sharing utensils (such as a cup or a glass), giving blood, skin-to-skin contact, or sitting on the same toilet seat as an infected person.

As yet, there is no effective vaccine against HIV, although a number of antiviral drugs can delay the onset of AIDS.

The transmission of AIDS can be controlled by a number of measures. These include:

- use of condoms – although transmission can still occur, its incidence is greatly reduced

- drug users not sharing needles, and using only sterile needles

- limiting the number of sexual partners.

Typhoid

Typhoid or typhoid fever is caused by a bacterium called *Salmonella typhi*. It is transmitted by drinking water contaminated with human faeces, or by flies transferring the bacterium from faeces to food. Before the twentieth century, typhoid was endemic throughout the world and typhoid epidemics were common. As a result of modern sanitation and sewage systems, it is now rare in the developed world. But in developing countries, where there is poor sanitation, it still causes illness and death to thousands of people. The World Health Organization estimates that there are over 20 million cases of typhoid every year, and over 200 000 deaths from the disease.

Typhoid outbreaks often occur in the slum areas of 'megacities' in Africa, South America or the far East, where sanitation is poor or non-existent (Figure 13.15). Typhoid epidemics also happen when a country experiences a disaster such as war, famine, earthquake or floods. These result in people living in temporary camps, where the crowded conditions and lack of sanitation allow the bacterium to spread.

The incubation period of the disease is about 2 weeks. It starts with 'flu-like symptoms: a high fever, headaches, a cough and a general feeling of being unwell. The disease develops over the following weeks, when the patient suffers stomach cramps, constipation or diarrhoea, vomiting and delirium (mental confusion). Diarrhoea leads to severe dehydration. About 10–20% of sufferers develop life-threatening symptoms, when the bacteria attack the wall of the gut, causing bleeding and even perforation of the gut. Toxins released by the bacteria can cause inflammation of the heart and multiple organ failure.

Vaccines against typhoid are available, and antibiotics such as penicillin are effective against the bacteria. Of course, these are often not available to the poor people most likely to suffer from the disease. **Oral rehydration therapy** (see Chapter 8, page 113) is very effective in counteracting the effects of the dehydration caused by diarrhoea.

Good sanitation and hygiene are essential in preventing the disease. It is only spread in places where human faeces or urine come into contact with food or drinking water. Unfortunately, this is likely to continue when so many people of the world continue to live in overcrowded, unhygienic conditions.

Tuberculosis

Tuberculosis (TB) is one of the most common human diseases, with millions of sufferers around the world. Nearly 2 million people die every year from TB. As with typhoid, the disease is much more common in developing countries, and is relatively rare in developed countries such as Western Europe and the United States.

There are several forms of TB, the most common being due to a bacterium called *Mycobacterium tuberculosis*. This type of TB is spread from person to person by droplet infection when an infected person coughs, sneezes or spits, so is more likely to spread in overcrowded, unhygienic living conditions.

Most people infected with the bacterium do not show any symptoms; only about one in ten people go on to get the active disease which, if untreated, kills over half of its victims. In some African or Asian countries, up to 80% of people test positive

Effective sanitation, which disposes of human waste in a way that ensures infections cannot be transmitted from person to person, is essential in preventing the spread of many diseases, as well as reducing pollution of rivers and other bodies of water. Methods of sewage treatment are described in Chapter 14, page 209.

Megacities are cities with populations of over 10 million people, such as Mexico City, Mumbai, Karachi or Buenos Aires. Many megacities are surrounded by shanty towns, where poor people live in overcrowded and unsanitary conditions, and disease is rife. One-sixth of the world's population now lives in these conditions.

Figure 13.15 *Typhoid outbreaks are common in many parts of the developing world, especially in the slum areas of 'megacities' such as Mumbai.*

for the bacterium, compared with only about 5–10% of people in a developed country such as the United States. Up to a third of the world's population is thought to be infected with TB.

TB usually infects the respiratory system, where the bacteria damage the lung tissue, forming swellings called tubercles, which gives the disease its name. The damaged lung tissue can be seen by X-ray (Figure 13.16). However, the bacteria can go on to infect other tissues and organs, such as the bones and lymphatic system. The bacteria also suppress the immune system, so that the patient is less able to overcome the disease.

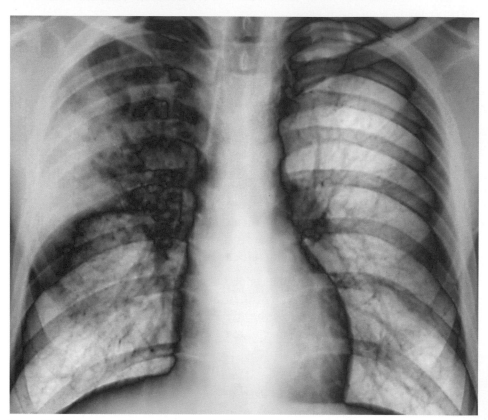

Figure 13.16 *A coloured X-ray of the chest of a TB patient. The lungs (white) contain affected areas showing as grainy dark blue patches (tubercles).*

Many people think that **chronic** means 'very bad' – it doesn't. Its correct meaning is 'long-term'. The opposite is **acute**, which means 'short-term'.

The symptoms of TB are a chronic cough, with fever, night sweats and weight loss. The patient may cough up blood.

It is very difficult to treat TB, involving long-term use of multiple antibiotics. It can take up to 18 months to kill all the bacteria. In addition, resistance of the bacteria to antibiotics is becoming a major problem (see Chapter 10, page 145). Treatment is also expensive, so is not an option for most people with the disease.

The most effective way of preventing the spread of TB would be to improve the standard of living in countries where TB is endemic. Less crowded housing and workplaces would reduce the opportunity for the bacterium to be passed from person to person, and if people were well fed and generally healthier, they would be less likely to develop the active form of the disease. Improvements like this are the reason why TB cases have been dramatically reduced in developed countries over the past 100 years. But, as with so many other diseases, it is unlikely that living standards in poor countries will improve enough to bring about this change in the near future.

An alternative method of control is vaccination. One effective method is the BCG vaccine.

BCG only really works well in preventing childhood TB. As yet, there is no reliable vaccine that protects adults, although several new vaccines are being developed. Vaccination programmes are under way in countries where TB is common in an attempt to eradicate the disease. For example, in South Africa, which has the highest incidence of TB in the world, all children under 3 years old are being vaccinated with BCG, and new vaccines for adults are under trial.

BCG stands for Bacille Calmette-Guérin, named after two French scientists who developed the vaccine.

Gonorrhoea

Gonorrhoea is one of several **sexually transmitted diseases** (STDs). It is common throughout the world. Gonorrhoea is caused by the bacterium *Neisseria gonorrhoeae*, which is passed from person to person during sexual intercourse. Symptoms usually appear a few days after infection, although about half of all infected people, mainly women, show no symptoms at all. Most men show symptoms, including discharge of pus from the penis and pain on urinating. There are also general symptoms, such as fever and headaches. Some women have a discharge from the vagina, or bleeding between menstrual periods. If the disease is untreated it becomes much more serious, with the infection spreading to the uterus, oviducts and ovaries, and can lead to a woman becoming infertile.

It is possible to cure gonorrhoea using antibiotics. Often a person with gonorrhoea will be infected with other bacterial STDs, so combinations of antibiotics are usually prescribed. The main method of prevention is to avoid sexual intercourse with people who might have the infection, or to use a condom. Condoms are 99% effective in preventing transmission of the disease. As with so many bacterial diseases, the bacterium has started to evolve resistance to some antibiotics. The World Health Organization is worried that antibiotic resistance will soon make it very difficult to treat the disease.

Fungal infections – thrush and athlete's foot

Few human diseases are caused by fungi. Those fungi that are pathogenic mostly cause infections of the skin or nails, or the moist mucous membranes of the mouth or vagina. One common fungal disease is thrush, caused by the fungus *Candida albicans*. Most people have this fungus living on their skin, and normally it causes no problems. However, sometimes a more serious thrush infection can develop. A common reason for this is if the person is taking a course of antibiotics for a bacterial problem. Our normal skin bacteria compete with the fungus, keeping it in check. The antibiotic may destroy these bacteria, so that the *Candida* is able to grow out of control. Thrush infections are common in babies and young children, as well as in adults (Figure 13.17).

Another disease caused by a fungus is **athlete's foot**. This usually grows on the warm, moist skin between the toes. Fungal spores may be transferred to the skin from the air or floor – a common place where the infection is picked up is in sports changing rooms. The fungus feeds on the outer layers of the skin. A bad infection causes sore, raw patches that may become infected by other organisms.

Fungal diseases are easily treated using antifungal drugs, either applied to the skin or taken by mouth.

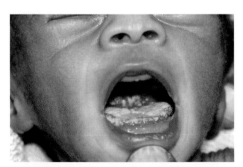

Figure 13.17 *Babies have a weaker immune system than most adults and sometimes develop thrush infections.*

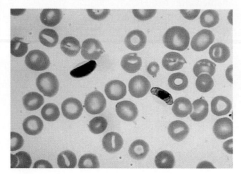

Figure 13.18 Plasmodium *parasites in human blood.*

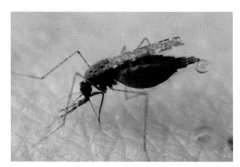

Figure 13.19 *A female* Anopheles *mosquito sucking human blood.*

Malaria

Malaria is caused by five different species of a protozoan pathogen called *Plasmodium* (Figure 13.18), which is passed to humans via an insect host – the mosquito.

The female *Anopheles* mosquito transmits *Plasmodium* in her saliva when she sucks blood from humans (Figure 13.19).

Figure 13.20 shows the life cycle of the *Anopheles* mosquito and Figure 13.21 shows the life cycle of the malarial parasite.

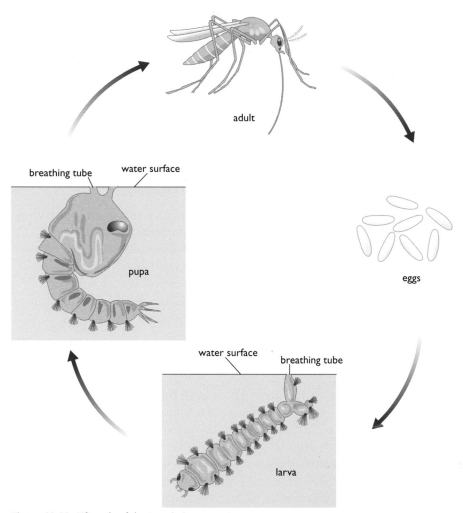

Figure 13.20 *Life cycle of the* Anopheles *mosquito.*

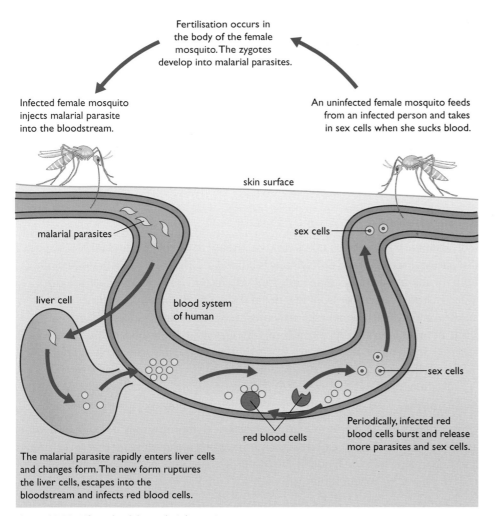

Fertilisation occurs in the body of the female mosquito. The zygotes develop into malarial parasites.

Infected female mosquito injects malarial parasite into the bloodstream.

An uninfected female mosquito feeds from an infected person and takes in sex cells when she sucks blood.

skin surface

malarial parasites

sex cells

liver cell

blood system of human

sex cells

Periodically, infected red blood cells burst and release more parasites and sex cells.

red blood cells

The malarial parasite rapidly enters liver cells and changes form. The new form ruptures the liver cells, escapes into the bloodstream and infects red blood cells.

Figure 13.21 *Life cycle of the malarial parasite.*

Malaria is widespread in tropical and sub-tropical parts of the world, and is estimated to cause the death of several million people every year, as well as making hundreds of millions so ill that they cannot work – contributing to poverty in the developing countries where it is endemic.

The malaria parasite spends anywhere between 2 weeks and several months in the person's liver before the next stage in the life cycle infects red blood cells. The well-known symptoms of malaria then appear – alternating cold sweats and fever, vomiting, joint pains and anaemia. Severe malaria causes coma and death, especially in young children.

If the life cycle of either the malarial parasite or the mosquito host can be broken at any point, then the transmission of malaria can be controlled. Controlling numbers of the mosquito means there will be fewer insects to transmit the protozoan pathogen. Controlling numbers of the *Plasmodium* parasite means there will be fewer opportunities for mosquitoes to take in the parasite and transmit it to other humans.

Control measures that have been tried include:

- the use of insecticides to kill the adult mosquitoes

- draining swamps that form the natural habitat of the larvae of the mosquitoes

- the use of drugs to target the various stages of the protozoan's life cycle

- stocking ponds with a fish called *Tilapia*, which feeds on the larvae of mosquitoes

- the use of insect repellents, wearing long-sleeved shirts and sleeping under mosquito nets to prevent bites from the adult mosquitoes.

Research is currently under way to develop a vaccine against *Plasmodium*.

Malaria is a difficult disease for our immune system to attack, because the protozoan parasite spends much of its time 'hidden' inside red blood cells, and changes form several times inside the body.

Schistosomiasis

Schistosomiasis (also called bilharzia) is a disease caused by several different species of a parasitic worm called *Schistosoma* (Figure 13.22). Although it is not usually fatal, the disease damages many internal organs and is debilitating – it prevents sufferers from working or leading a normal life. The disease is sometimes called 'snail fever' because the parasite is transmitted to humans via freshwater snails. It is common in many tropical parts of Africa, Asia, the Caribbean and South America, in areas where there are snails that carry the parasite. It often infects children who swim in rivers or lakes where the snails live. The World Health Organization estimates that over 200 million people are infected worldwide, mostly in areas with poor sanitation and no access to clean drinking water.

Schistosomiasis is a long-term (chronic) disease. Symptoms may be mild, although sufferers from the disease can develop more serious symptoms, such as fever, chills, diarrhoea, skin rash and blood in the urine. Some organs, such as the liver, spleen and lymph nodes, become enlarged.

Larvae of the worms are released by freshwater snails. The larvae swim in the water and penetrate the skin of people in the water. In the body, the larvae develop into adult worms, which live inside blood vessels in several different organs, including the liver, intestines and bladder, where they feed on red blood cells. They mate and release eggs, which pass out in faeces or urine to reinfect more snails.

There are four main ways to control schistosomiasis:

- treatment with drugs to kill the worms in the body

- killing the snails to interrupt the life cycle

- improving sanitation

- health education.

An effective drug is available that will kill the worms. But controlling the freshwater snails is more effective in preventing the spread of the disease. This is done with chemicals or by introducing natural predators, such as crayfish. Improved sanitation also allows the cycle to be broken – if human faeces and urine do not enter rivers and lakes, this prevents the worms' eggs from reinfecting more snails. Health education is also used to inform villagers about the dangers of entering water where the snails live.

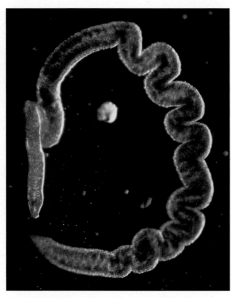

Figure 13.22 *The parasitic worm that causes schistosomiasis.*

Defence against disease

Immunity

General immune responses are aimed at any invading microorganism. The most important of these is **phagocytosis**. Some white blood cells can ingest (take in) microorganisms. They do this by forming extensions called **pseudopodia** that enclose the microorganism. This is called phagocytosis and the white blood cells that carry it out are **phagocytes**. Once ingested, the phagocyte encloses the microorganism in a vacuole and digests it. Figure 13.23 shows a phagocyte ingesting a yeast cell.

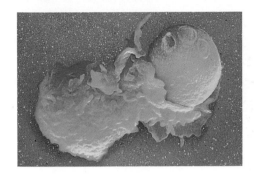

Figure 13.23 *A phagocyte ingesting a yeast cell.*

Specific immune responses are those that our body makes to each individual disease-causing organism. There is one response to the TB bacterium, a different response to the common cold virus, and so on. To make these individual responses, our body has first to be able to recognise the different microorganisms. The cells in our body that can do this are called lymphocytes. There are many kinds of **lymphocytes**, but two are particularly important – **B-lymphocytes** and **T-lymphocytes**.

Microorganisms have individual 'marker chemicals' or **antigens** on their surface. Lymphocytes have **receptor proteins** on their surface that can bind with the antigens on the surface of the microorganisms. Each lymphocyte has slightly different receptor proteins from other lymphocytes and so can bind with different antigens and recognise a different microorganism.

When a lymphocyte binds with an antigen, it becomes activated and starts to divide rapidly. This will eventually produce millions of the same type of lymphocyte, capable of recognising the same type of microorganism. So far, the story is the same for B-lymphocytes and T-lymphocytes. What happens as the lymphocytes multiply, however, is different.

When B-lymphocytes are activated, most start to produce **antibodies** that are specific to the antigens on the surface of the microorganism. These antibodies bind with the antigens and cause the microorganisms to 'clump' together (Figure 13.24). In this form they are inactive and can be easily killed by phagocytes. Sometimes antibodies cause the cells to burst open.

A phagocyte isn't a particular type of white blood cell. There are several, quite different, types of white blood cell that are phagocytes. The term describes a common function of these cells. Amoebae are also phagocytic!

Lymphocytes are white blood cells that are produced from actively dividing stem cells in the bone marrow. B-lymphocytes complete their development in the spleen and the lymph nodes. T-lymphocytes complete their development in the thymus gland – just above the heart.

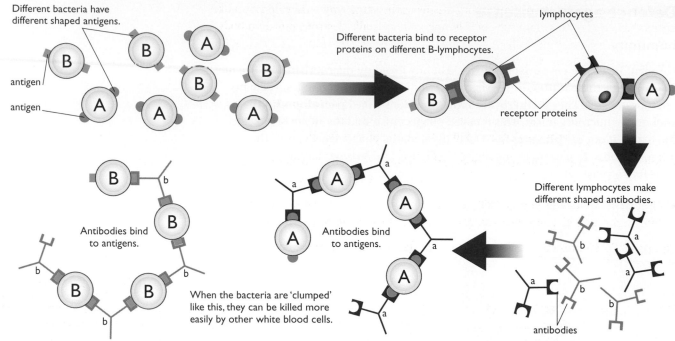

Figure 13.24 *How B-lymphocytes produce antibodies and destroy microorganisms.*

Antibodies are 'Y' shaped protein molecules that bind to a specific protein (antigen) on the surface of a microorganism.

this part of the molecule is specific to one antigen

this part of the molecule is the same in all antibodies

Figure 13.25 *The structure of an antibody molecule.*

It is the *memory cells* that remain in the blood for long periods of time, *not* the antibodies.

Some of the activated B-lymphocytes do not get involved in killing microorganisms at this stage. Instead, they develop into **memory cells**. Memory cells make us **immune** to a disease. These cells remain in the blood for many years, in some cases a lifetime. If the same microorganism re-infects, the memory B-lymphocytes start to reproduce and produce antibodies. This secondary immune response is much faster and more effective than the first (primary) response. The antibody level quickly becomes much higher and the microorganisms are killed before they have had time to multiply to the level where they would cause disease. Figure 13.26 illustrates this.

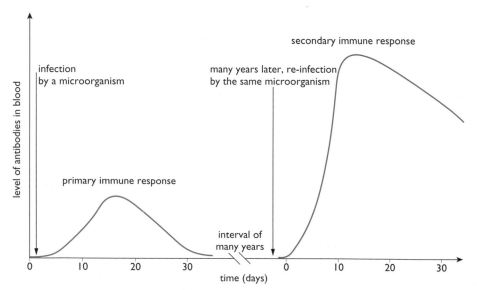

Figure 13.26 *A graph showing the levels of antibody production in the primary and secondary immune responses. Many years may pass after the primary response before we are re-infected and need to make a secondary response.*

T-lymphocytes do not destroy microorganisms in the same way – they do not make antibodies. In fact, the main type of T-lymphocyte actually destroys our own body cells. If this doesn't immediately make any kind of sense, remember that cancer cells are our own body cells 'gone wrong'. viruses also enter our cells and reproduce inside them – here antibodies cannot reach them.

T-lymphocytes are able to recognise virus-infected cells (and sometimes cancerous cells as well) because of telltale marker antigens on the surface of the cell. The T-lymphocytes target these cells and kill them in one of two ways:

- Some T-lymphocytes release chemicals that 'punch a hole' in the body cell. All the contents leak out, the cell dies and with it, the viruses inside.

- Other T-lymphocytes press a 'self-destruct button' in the cell. They activate a 'programmed cell death' process that is part of the genetic code of every cell.

Figure 13.27 summarises how T-lymphocytes destroy virus-infected cells.

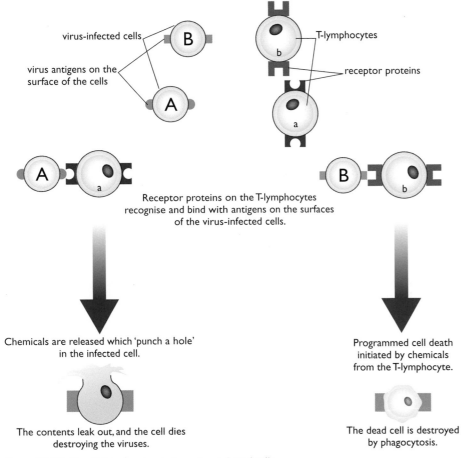

Figure 13.27 *How T-lymphocytes destroy virus-infected cells.*

Our immune systems are at their most active in childhood and adolescence. This is the period of our lives when we meet most diseases for the first time. Although we retain an active immune system throughout adulthood, it is less effective at combating new diseases than it was in our teens.

As with B-lymphocytes, some of the T-lymphocytes become memory cells and remain in the body for many years. They can also mount a speedy secondary immune response if the same microorganism re-infects us. Again, the memory T-lymphocytes are the basis of our immunity to disease.

The process of becoming immune to a disease is called **immunisation**. The responses described so far are **natural active** responses made by *our own bodies* to make us immune to a disease. There are other types of immunity in addition to these.

Natural passive immunity

Passive immunity is a means of acquiring immunity without having to actually produce an immune response ourselves. Normally this occurs only twice in our lives, both times when we are very young:

- We receive antibodies across the placenta before we are born. The antibodies cross the placenta from the mother's blood in increasing amounts throughout pregnancy.

- We also receive antibodies from our mothers in colostrum and breast milk. Colostrum is the liquid produced for the first few days after birth, before the mother produces breast milk proper. Both contain antibodies, but colostrum is a particularly rich source.

Passive immunity is short lived. Because babies haven't made the antibodies themselves, the levels drop once breastfeeding stops. However, this passive immunity protects newborn babies for a vital few months while their own immune systems are developing.

Artificial active immunity – vaccination

Vaccination is a way of giving nature a helping hand (Figure 13.28). We can be made immune to a disease without actually contracting (having) the disease itself. A person is injected with some 'agent' that carries the same antigens as a specific disease-causing microorganism. Lymphocytes 'recognise' the antigens and multiply exactly as if that microorganism had entered the bloodstream. They form memory cells and make us immune to the disease. The actual vaccine that is injected may be:

- an **attenuated** (weakened) strain of the actual microorganism (e.g. polio, TB (tuberculosis) and measles vaccines)

- dead microorganisms (e.g. whooping cough and typhoid fever vaccines)

- modified toxins of the bacteria (e.g. the toxins of the tetanus and diphtheria bacteria)

- just the antigens (e.g. the influenza vaccine)

- harmless bacteria, genetically engineered to carry the antigens of a different disease-causing microorganism (e.g. the hepatitis B vaccine).

Different diseases are endemic in different parts of the world. Travellers should seek advice from their doctor before visiting foreign countries about any vaccinations that may be necessary.

'Breast is best' is a slogan used to encourage natural breastfeeding rather than bottle-feeding of babies. One reason why breast is best is that only breast milk can give immunity because of the antibodies it contains.

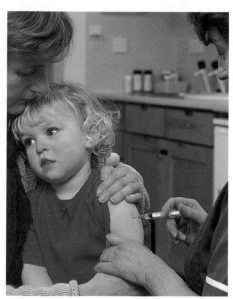

Figure 13.28 *A vaccine is injected to stimulate an immune response to a particular disease-causing organism.*

Edward Jenner performed the first vaccination in 1796. At that time, smallpox was a major killer in England. Jenner was an English country doctor and had noticed that people who caught cowpox (a similar, but much milder disease) hardly ever caught smallpox. Jenner vaccinated a small boy called Edward Phipps to show that cowpox could protect against smallpox (Figure 13.29). He first infected the boy with cowpox and allowed him 2 months to recover from the disease. He then infected him with pus from a smallpox sufferer.

Edward Phipps did not contract smallpox and we now understand exactly why – and so should you. The antigens on the cowpox virus are the same as those on the smallpox virus. They stimulate exactly the same immune response and the memory cells formed give immunity against smallpox. The word *vaccination* comes from the Latin word *vacca*, meaning a cow.

Vaccination has reduced the incidence of many serious diseases and has virtually eradicated smallpox from the planet. However, there may be risks with vaccination. If there is a microorganism in the vaccine, it may recover the ability to become infective. In the 1950s, a batch of polio vaccine was not properly prepared, with the result that many children actually contracted polio, instead of being protected against it. Today, the risks are better understood and procedures are much more effective.

Recently there has been concern that the triple vaccine MMR (for mumps, measles and rubella) may cause autism in a small number of cases. The position is far from clear, but the most recent research suggests that there is no link between MMR and autism. However, the problems have raised many questions about vaccination.

Sometimes, vaccination against a microorganism is virtually impossible because the microorganism mutates rapidly with some of the mutations resulting in different surface antigens. This means that any memory cells the body has made will not recognise the new form of the microorganism. A new vaccine must be produced that will stimulate the production of new memory cells. This is why we get so many colds. The common cold virus has a very high mutation rate so by the time a vaccine has been produced, the virus has mutated again.

The influenza ('flu) virus also mutates regularly, but not as fast as the common cold virus. There is enough time to produce a 'flu vaccine that will be effective before the virus mutates again.

Artificial passive immunity

Sometimes people need help immediately so it is not appropriate to give them a vaccination that would make them immune to a disease sometime in the future. In these cases a person can be injected with the actual antibodies to help to boost their own immune response. This is often done in the case of rabies, although there is also a vaccination against the disease.

Figure 13.29 *A statue of Edward Jenner infecting Edward Phipps with a vaccine against smallpox.*

Louis Pasteur and rabies

Louis Pasteur is most commonly remembered for his work on prevention of spoilage of foods, and for the process that bears his name – Pasteurisation. However, he also did important work in the field of vaccination and developed a vaccine against rabies. Rabies is a viral disease, common in animals, that can be transmitted to humans through a bite. Most cases of rabies in humans are transmitted from rabid dogs (dogs with rabies). Figure 13.30 shows the distribution of rabies throughout the world, and the animals mainly responsible for the transmission of rabies in the different areas.

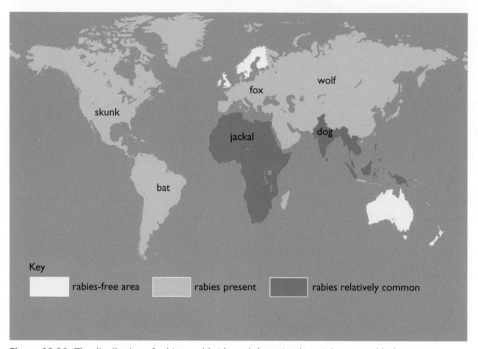

Figure 13.30 *The distribution of rabies worldwide and the animals mainly responsible for its transmission.*

> The virulence of a microorganism is its ability to infect and cause disease.

Pasteur had read about Jenner's work and had found similar cases of mild forms of a disease protecting against a more severe form. More importantly, he developed a method of making bacteria lose their virulence. Pasteur found that by keeping cultures of bacteria for a long time – by taking samples and culturing the samples and then taking more samples and culturing them and so on – the bacteria lost some of their virulence. He was able to do this for anthrax and rabies – the first rabies vaccine had been produced. The vaccine had extremely unpleasant side effects and was far from safe, but his work represented a significant breakthrough in immunology. Today's anti-rabies vaccines are much safer and have fewer side effects.

Antibiotics

The discovery of penicillin

In 1928, in St Mary's Hospital in London, a research microbiologist noticed that one of his cultures of bacteria had become contaminated. A mould was also growing on the culture. He looked more carefully and noticed that there were no bacteria around the area where the mould was growing. It probably looked something like the Petri dish shown in Figure 13.31(a).

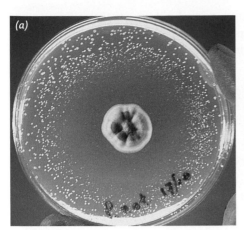

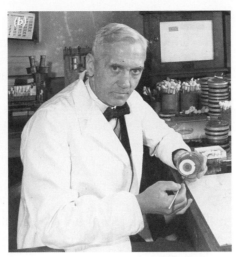

Centuries before antibiotics were discovered, people living in the countryside knew that rubbing certain moulds into wounds stopped them from becoming infected. They were, unknowingly, administering the antibiotic penicillin that was being produced by the mould.

Figure 13.31 *(a) A Petri dish with a fungus and bacterial colonies growing on the agar. The fungus is secreting an antibiotic that is inhibiting the growth of the bacteria. (b) Alexander Fleming (1881–1955).*

He then deliberately repeated the contamination of a culture – with the same results. He reasoned that the fungus must be secreting a substance that inhibited the growth of bacteria in some way. The researcher's name was Alexander Fleming and he named the substance penicillin – after the mould *Penicillium* that contaminated his cultures. The first antibiotic had been discovered.

Fleming never extracted the pure penicillin. This was left to other researchers – Howard Florey and Ernst Chain. They isolated and purified penicillin and were able to confirm that penicillin did have an antibacterial effect. Florey and Chain also began to develop techniques to scale up the production of penicillin.

Penicillin was the first antibiotic to be discovered. When it was first used to treat bacterial diseases, there were spectacular results. The bacteria had no resistance to the drug and almost 'miracle cures' seemed possible. However, as penicillin was used more and more frequently, mutations in some bacteria gave them resistance to it. Natural selection allowed these bacteria to survive and reproduce. As these resistant strains became more widespread, so the effect of penicillin diminished.

Other antibiotics were discovered and used. These included drugs like streptomycin, chloramphenicol, aureomycin and many others. The pattern was the same with these drugs. Initially they produced spectacular results, then resistant strains appeared, became widespread and the effect diminished.

As people became more aware of antibiotics, they expected (and often demanded) to be prescribed them for even the most trivial infections. This led to over-prescription of the drugs and rapid emergence of resistant strains. Doctors began to prescribe antibiotics in combination for greater effect. Bacteria then evolved that were resistant to several antibiotics – so called 'super-bugs'.

Today, doctors are more aware of this problem and prescribe antibiotics more judiciously. Also, new antibiotics are synthesised in laboratories. There does not need to be a 'search' through a range of microorganisms to see if any produce an antibiotic so far undiscovered.

How antibiotics act

Bactericidal antibiotics actually kill bacteria, while **bacteriostatic** antibiotics just stop them from reproducing. **Broad spectrum** antibiotics act against a range of bacteria, while others are more limited in their target. Table 13.3 shows some of the ways in which antibiotics act.

Example	Mode of action	Effect	Type of antibiotic
penicillin	interferes with manufacture of bacterial cell wall	weakens cell wall; water enters by osmosis and bursts the cell (osmotic lysis)	bactericidal
nalidixic acid	interferes with DNA replication	bacteria are not killed, but cell division is impossible so they cannot multiply	bacteriostatic
tetracycline	interferes with protein synthesis	no enzymes can be made to control the reactions in the bacterial cells	bactericidal

Table 13.3: *Comparison of different types of antibiotics.*

Figure 13.32 shows how penicillin acts on bacterial cells. It is particularly effective against bacteria when they are dividing as it interferes with the manufacture of the new cell walls of the daughter cells.

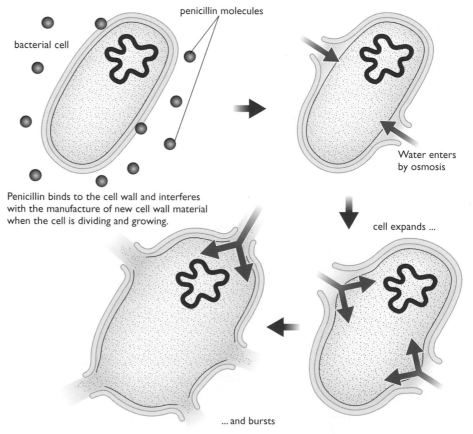

penicillin molecules

bacterial cell

Water enters by osmosis

Penicillin binds to the cell wall and interferes with the manufacture of new cell wall material when the cell is dividing and growing.

cell expands ...

... and bursts

Figure 13.32 *The way in which penicillin kills bacteria.*

Antibiotics have no effect at all on viruses for the following reasons:

- Viruses have no cell walls – so antibiotics like penicillin can have no effect.

- Viral DNA only replicates inside other cells once the virus has infected the cell.

- Viruses do not synthesise their own proteins; they instruct the cell they infect to synthesise their proteins.

- Viruses are only active *inside* other living cells and are protected by the cell as most antibiotics cannot enter human cells.

Producing penicillin

Penicillin is secreted by the fungus *Penicillium*. Different strains of the fungus can produce slightly different penicillins, that can be active against different bacteria. New strains of *Penicillium* are constantly 'screened' for antibiotic production. Any strain that appears to produce a new type of penicillin with significant anti-bacterial activity is then cultured for further studies to see if the fungus is suitable for bulk production of penicillin.

Biochemists can now produce 'designer' penicillins. This involves altering a basic penicillin molecule to give it a slightly different shape. It is done in one of two ways:

- Altering the 'diet' of the fungus during production – this can modify the side chain on the penicillin molecule (Figure 13.33).

- Some strains of the fungus produce penicillin with no side chain at all. A side chain can be created chemically that will bind with the desired cell wall material, and then added to the penicillin molecule.

These two techniques allow production of thousands of different, semi-synthetic penicillins. Each will have a slightly different shape and so have different antibiotic activity.

Food hygiene

Many diseases are transmitted through infected food. Humans can do a great deal, both collectively and individually, to minimise the risk of infection, by ensuring that food is hygienically prepared, stored and preserved. Many of the food hygiene measures we can carry out are aimed at blocking transmission routes.

We can prevent the transmission of food-borne microorganisms at various stages. We can:

- prevent microorganisms from getting into food in the first place

- treat food to minimise the rate at which microorganisms multiply within it (food preservation)

- cook food properly to kill any microorganisms present.

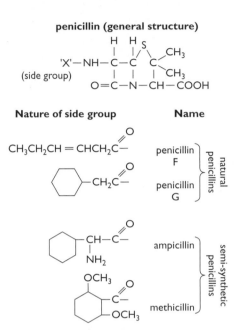

Figure 13.33 *Different side chains give a penicillin molecule different properties.*

Figure 13.34 *Contamination of food can be prevented during production by keeping animals in hygienic conditions.*

Preventing microorganisms from entering food

Microorganisms can enter food at several stages: at the point of production, at the point of sale, and during storage after production or in the home.

The conditions under which livestock are reared can affect their health and the food we get from them. For example, some cheeses are still made from untreated (unpasteurised) milk. Any harmful bacteria in the milk of the cow or goat could find their way into the cheese and from there to us. Poor housing of chickens can allow food-poisoning bacteria such as *Salmonella* to spread rapidly among the birds and lead to infection of the eggs (Figure 13.34).

The microorganisms that cause disease in crop plants rarely have significant effects on humans, but harvested crops such as cereals must be stored properly. Pests could enter the store and contaminate the food with excreta and disease-causing microorganisms.

At the point of sale, food should be stored so as to minimise transmission from humans. Packaging of food prevents this, but loose vegetables and fruit are more at risk. 'Display until' and 'best before' dates are important (Figure 13.35). They tell us when the food is likely to become unsafe to eat because of contamination by microorganisms or toxins from microorganisms.

Figure 13.35 *'Best before' dates help us to know if foods are likely to be contaminated.*

In the home, food should be stored in a way that minimises the likelihood of transmission. For example:

- cooked and raw foods should not be stored together (bacteria in uncooked food may be transferred to cooked food)

- foods that have been frozen should not be refrozen after cooking (any bacteria that remain will multiply when food is recooked)

- food should not be left open to the air on a work surface (bacteria in the air or carried by insects could contaminate it).

Food preservation

Some methods of food preservation, such as salting and pickling food, have been used for hundreds of years, while other treatments, such irradiation of food, are very recent methods. Table 13.4 describes some methods of food preservation.

Technique	Principles and procedures of the technique	Example of foods preserved this way	Possible drawbacks
salting	High salt concentrations make it impossible for bacteria to multiply. Bacterial cells lose water by osmosis and are killed. Some foods are covered in salt, others are soaked in brine (salt water).	some meats, fish (kippers), vegetables	Some bacteria can withstand very high salt concentrations.
pickling	Foods are bottled in vinegar – a weak, flavoured solution of ethanoic (acetic) acid. The low pH inactivates most microorganisms.	onions, red cabbage, herrings, mayonnaise	It alters the taste of the food.
pasteurisation	The food is heated to between 63 and 65 °C (Figure 26.7) for 30 minutes or 71.5 °C for 15 seconds. Both techniques kill pathogenic bacteria.	milk, cream, ice-cream, fruit juices, beer	It does not kill all bacteria – spoilage of the food is merely delayed. Spores of harmful bacteria may survive.
ultra-heat-treatment (UHT)	Superheated steam at temperatures of 135–160 °C is blown through the food for 2 seconds. This kills all bacteria and spores.	milk	It alters the taste of the milk.
canning	Food is packed in cans, heated to high temperatures, sealed and then reheated to temperatures of 105–160 °C. The high temperatures kill bacteria and spores.	beans, soup, tomatoes	Cans can be damaged when cooled and in transit. This can allow bacteria to enter the can and multiply. Beware dented cans!
drying	Drying removes water from the food so bacteria cannot digest and absorb it. Manufacturers dry food by blowing hot air through it.	cereals, grains, some fruits	Water can easily re-enter once the packets are opened. Bacteria can then cause spoilage.
freezing	Foods are cooled rapidly to temperatures of −10 °C. Rapid freezing prevents the formation of large ice crystals which could alter the texture and flavour of the food.	many vegetables, meats, prepared meals	none
irradiation	High-energy gamma radiation from sources such as radioactive cobalt or radioactive caesium is passed through the food. All bacteria and spores are killed, although toxins produced by the bacteria still remain. There is no change in the taste of the food.	Some countries permit irradiation of potatoes, onions, shellfish and some fruits.	Some people are concerned that food may be made radioactive or that carcinogens may be formed in the food (although neither is true).

Table 13.4: Common methods of food preservation.

End of Chapter Checklist

You should now be able to:

✓ recall the main types of pathogenic microorganisms, including bacteria, viruses, fungi and protozoa

✓ recall a brief description of the structure, nutrition and reproduction of bacteria

✓ recall a brief description of the structure and reproduction of viruses

✓ recall the general course of a disease, including methods of infection, incubation and signs and symptoms, with particular reference to the diseases listed below

✓ understand the difference between endemic and epidemic diseases

✓ recall the methods of transmission, treatment and prevention of spread of the viral diseases influenza, poliomyelitis and AIDS

✓ recall the methods of transmission, treatment and prevention of spread of the bacterial diseases typhoid, tuberculosis and gonorrhoea

✓ understand the relationship between the housefly as a vector in the transmission of the typhoid bacillus, and the treatment and prevention of spread of typhoid and its vector

✓ recall the methods of transmission, treatment and prevention of spread of the fungal diseases thrush and athlete's foot

✓ understand the relationship between the mosquito as a vector in the transmission of the malaria parasite, and the treatment and prevention of spread of malaria and its vector

✓ recall the nutrition and life cycle of the protozoan parasite *Schistosoma*

✓ recall the worldwide effects of the disease schistosomiasis (bilharzia), including methods of preventing its spread

✓ understand the meaning of immunity, and that immunity can be natural or artificial, and active or passive

✓ understand the antibody–antigen reaction

✓ explain what a vaccine is and how it works

✓ recall the sources and role of antibiotics

✓ understand the hygienic methods of food preparation, storage and preservation

Questions

1 *a)* Draw a diagram to show the structure of a typical virus particle.

   b) Is a virus a living organism? Explain your answer.

   c) Explain the statement 'viruses are all parasites'.

2 The diagram below shows the structure of a typical bacterium.

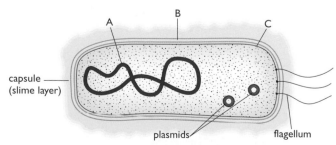

a) Identify structures A, B and C.

b) What is the function of the flagellum?

c) 'All bacteria are pathogens'. Briefly discuss whether this statement is true.

3 What is the difference between an infectious and a non-infectious disease? Give an example of each.

4 AIDS is one of the biggest worldwide killer diseases. It is caused by the human immunodeficiency virus (HIV).

a) Give two ways in which HIV can enter the body.

b) What does being 'HIV positive' mean?

c) The graph shows the changes in the numbers of helper T-lymphocytes during an AIDS infection.

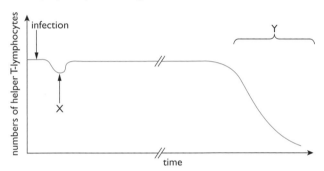

Suggest explanations for the changes in numbers at:

i)   the period marked X

ii) the period marked Y.

5 Diseases are sometimes endemic in a certain area of the world.

a) What does endemic mean?

b) What happens when an epidemic of a disease occurs?

c) Give one example of a disease that has undergone an epidemic.

d) Why should you make sure that you are up-to-date with your vaccines before travelling abroad?

6 Blood has several immune functions. Some of them are general, others are specific.

a) Name two general immune functions of blood.

b) Describe the role of T-lymphocytes in combating a viral infection.

c) Describe how memory T-lymphocytes are formed and explain their importance in giving lasting immunity to a disease.

7 The diagram shows the levels of antibodies produced by a person during an initial infection and later, when re-infected by the same microorganism.

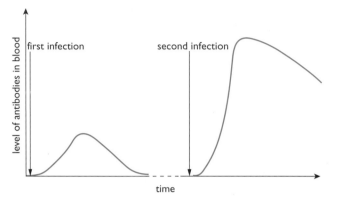

a) Give the names of the initial and subsequent responses by the body to infection.

b) Describe the differences between the two responses in relation to level, speed and duration.

c) Explain how the initial response is produced.

8 Gonorrhoea is a sexually transmitted disease caused by a bacterium. The graph shows changes in the incidence of gonorrhoea from 1925–1990.

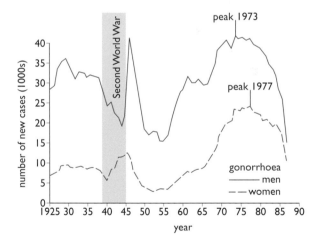

a) Name two other sexually transmitted diseases.

b) Describe the general trends in the numbers of cases of gonorrhoea from 1925 to 1990.

c) Suggest reasons for the increases in the number of cases of gonorrhoea:

i)   during the 1960s

ii) during the early 1970s.

d) Suggest how the increase in AIDS cases in the 1980s may be linked to the decrease in the incidence of gonorrhoea.

e) Gonorrhoea can usually be cured by giving penicillin.

   i) Describe how penicillin would kill the bacteria causing gonorrhoea.

   ii) Suggest why penicillin is not always effective in the treatment of gonorrhoea.

9 *a)* List four ways in which infectious diseases can be transmitted. In each case, give an example of a disease transmitted in that way.

  b) Give two examples of insects that act as vectors for disease. In each case, explain how the insect is responsible for the spread of the disease.

  c) For each of the two insect vectors you named in (b), describe two measures that can be taken with the insect to control the spread of the disease. Explain how each method works.

10 *a)* Explain the principles behind the following methods of food preservation:

   i) salting

   ii) pasteurisation

   iii) pickling.

  b) Why should food not be consumed after the 'use by' date?

Chapter 14: Human Influences on the Environment

Ecosystems

Humans inhabit many of the Earth's ecosystems. An ecosystem can be defined as a distinct, self-supporting system of organisms interacting with each other and with their physical environment. It can be small, such as a wood; or large, like a tropical rainforest.

Whatever their size, ecosystems usually have the same components:

- **producers** – green plants, that photosynthesise to produce food

- **consumers** – animals, including humans, that eat plants or other animals

- **decomposers** – bacteria and fungi that break down dead material and help to recycle nutrients

- the **physical environment** – the non-biological components of the ecosystem, such as water, soil and air.

Plants and photosynthesis

Plants use the simple inorganic molecules carbon dioxide and water, in the presence of chlorophyll and light, to make glucose and oxygen. This process is called **photosynthesis**.

It is summarised by the equation:

$$\text{carbon dioxide} + \text{water} \xrightarrow[\text{chlorophyll}]{\text{light}} \text{glucose} + \text{oxygen}$$

or:

$$6CO_2 + 6H_2O \longrightarrow C_6H_{12}O_6 + 6O_2$$

The role of the green pigment, chlorophyll, is to absorb the light energy needed for the reaction to take place. The products of the reaction (glucose and oxygen) contain more energy than the carbon dioxide and water. In other words, photosynthesis converts light energy into chemical energy.

You will probably have noticed that the equation for photosynthesis is the reverse of the one for aerobic respiration (see Chapter 1):

$$C_6H_{12}O_6 + 6O_2 \longrightarrow 6CO_2 + 6H_2O \text{ (plus energy)}$$

Respiration, which is carried out by both animals and plants, *releases* energy (but not as light) from the breakdown of glucose. The chemical energy in the glucose came originally from light 'trapped' by the process of photosynthesis.

Plants can convert glucose into many other organic molecules. The first product from glucose is usually starch, which is stored in the plant's leaves, roots or other organs. Many of the staple foods that we eat are rich in starch (Figure 14.1).

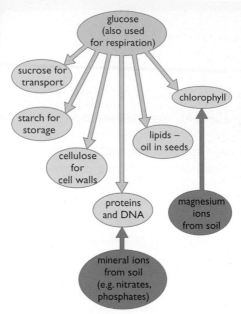

Figure 14.2 *Compounds that plant cells can make from glucose.*

It is not only plants that make food. Some simple organisms, such as algae and some species of bacteria, are able to photosynthesise, and are the main source of food and oxygen in the sea and lakes.

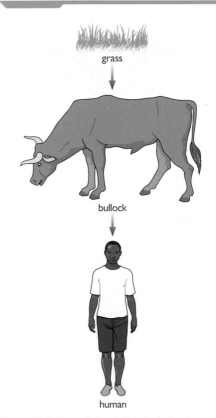

Figure 14.3 *A simple food chain including humans.*

A bullock is a young bull, bred for its meat.

Figure 14.1 *All these foods are made by plants and contain starch.*

Some glucose is converted into other sugars, such as fructose (fruit sugar) or sucrose (table sugar). Much glucose is used by the plants to make the cellulose of their cell walls.

All these compounds are carbohydrates. Plant cells can also convert glucose into lipids (fats and oils), which are needed to make the membranes of cells, and are also stored in fruits and seeds such as peanuts, sunflower seeds and olives.

Carbohydrates and lipids both contain only three elements – carbon, hydrogen and oxygen –so plants can make these from glucose without the need for a supply of other elements. Proteins contain these elements too, but all amino acids (the building blocks of proteins) also contain nitrogen. Plants obtain this from the soil as nitrate ions, along with other minerals, such as magnesium to make chlorophyll, and phosphate to make DNA and ATP. Some of the products that a plant makes from glucose are summarised in Figure 14.2.

So plants are the source of all the food that animals, including humans, eat. Through photosynthesis, plants also make the oxygen that aerobic animals need to respire.

Feeding relationships

The simplest way of showing feeding relationships within an ecosystem is a **food chain** (Figure 14.3).

In any food chain, the arrow ($\rightarrow$) means 'is eaten by'. In the food chain shown, the grass is the **producer**. It is a plant, so it can photosynthesise and produce food materials. The bullock is the **primary consumer**. It is an animal, that eats the producer. Since it eats plants it is also a **herbivore**. The human in this chain is the **secondary consumer** and **carnivore**. The different stages in a food chain (producer, primary consumer and secondary consumer) are called **trophic levels**.

Many food chains have more than three links in them. For example:

filamentous algae → mayfly nymph → caddis fly larvae → salmon → human

In this freshwater food chain, the extra links in the chain mean that the salmon is a **tertiary consumer**, and the human a **quaternary consumer**. The carnivore at the end of a food chain (in this case the human) is known as the **top carnivore**.

Most food chains involving humans are very short. If someone is a vegetarian, the chain only involves two organisms, the plant that made the food and the person that eats it. If a person eats meat or fish, the food chain is longer (Figure 14.3).

Food chains are a convenient way of showing the feeding relationships between a few organisms in an ecosystem, but they oversimplify the situation. The 'grass → bullock → human' food chain above implies that only the bullock feeds on grass, which is not true. Many small herbivorous animals may also feed on grass, such as insects, snails and mice. A clearer picture of the feeding relationships in an ecosystem is gained by linking several food chains together into what is called a **food web** (Figure 14.4).

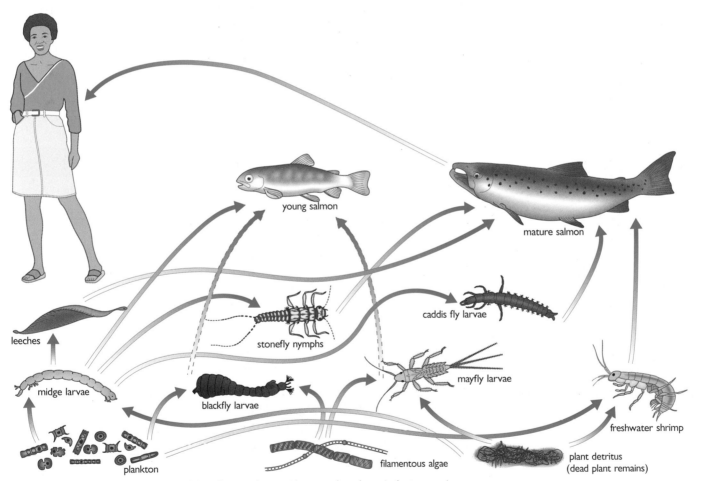

young salmon

mature salmon

caddis fly larvae

leeches

stonefly nymphs

mayfly larvae

midge larvae

blackfly larvae

freshwater shrimp

plankton

filamentous algae

plant detritus (dead plant remains)

Figure 14.4 *The freshwater food web of the salmon. A human that eats the salmon is the top carnivore.*

Although food webs give us more information than food chains, they do not give any information about how many organisms are involved, or their mass. Nor do they show the role of the decomposers. To see this, we must look at other ways of presenting information about feeding relationships in an ecosystem.

Ecological pyramids

Ecological pyramids are diagrams that represent the relative amounts of organisms at each trophic level in a food chain. There are two main types:

- **pyramids of numbers**, which represent the numbers of organisms in each trophic level in a food chain, irrespective of their mass

- **pyramids of biomass**, which show the total mass of the organisms in each trophic level, irrespective of their numbers.

Consider these two food chains:

grass → grasshopper → frog → bird

oak tree → aphid → ladybird → bird

Figures 14.5 and 14.6 show the pyramids of numbers and biomass for these two food chains. The two pyramids for the 'grass' food chain look the same. The numbers at each trophic level decrease. The total biomass also decreases along the food chain – the mass of all the grass plants in a large field would be more than that of all the grasshoppers, which would be more than that of all the frogs, and so on.

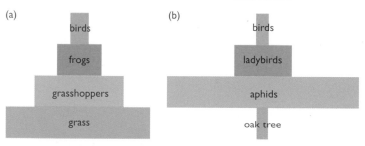

Figure 14.5 *Pyramids of numbers for the two food chains.*

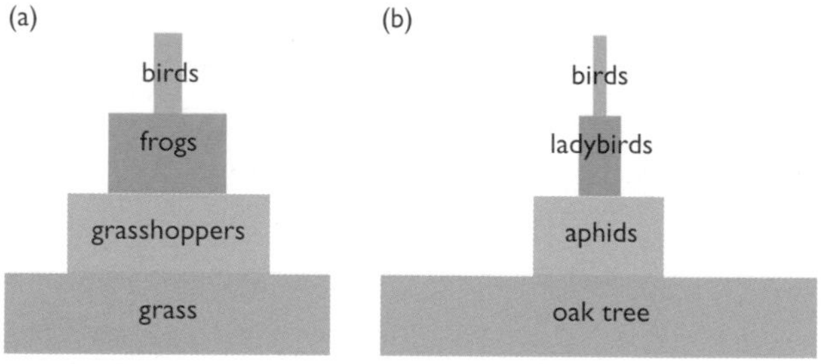

Figure 14.6 *Pyramids of biomass for the two food chains.*

The two pyramids for the 'oak tree' food chain look different because of the size of the oak trees. Each oak tree can support many thousands of aphids, so the numbers increase from first to second trophic levels. But each ladybird will need to eat many aphids and each bird will need to eat many ladybirds, so the numbers decrease at the third and fourth trophic levels. However, the *total* biomass decreases at each trophic level – the biomass of one oak tree is much greater than that of the thousands of aphids it supports. The total biomass of all these aphids is greater than that of the ladybirds, which is greater than that of the birds.

Why are diagrams of feeding relationships pyramid shaped?

The explanation for this involves energy losses. When a bullock eats grass, not all the materials in the grass end up as bullock. There are energy losses:

- Some parts of the grass are not digested, and so are not absorbed (even though a bullock has a very efficient digestive system) – they pass out in the animal's faeces.

- Some of the materials absorbed from the digested food form excretory products, such as urea.

- Much of the material from the grass is respired to release energy, with the loss of carbon dioxide and water.

Only a small fraction of the material in the grass ends up in new cells in the bullock. This is shown in Figure 14.7. The values in the boxes show the fate of energy in the food eaten by a bullock from 1 m^2 of grass per year. The units are kilojoules (kJ). Note how little energy from the grass ends up in the new tissue of the bullock – about 4%. In other words, only 4% of the energy in the producer (grass) is transferred to the primary consumer (bullock).

This does not just apply to this food chain. Generally, less than 10% of energy is transferred from one trophic level to the next in all food chains. This means that smaller and smaller amounts of biomass are available for growth at successive trophic levels. The shape of pyramids of biomass reflects this.

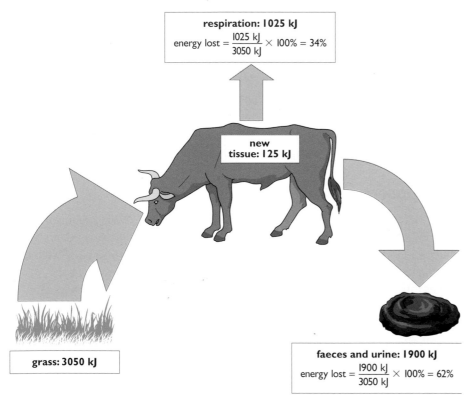

respiration: 1025 kJ

energy lost $= \dfrac{1025\ kJ}{3050\ kJ} \times 100\% = 34\%$

new tissue: 125 kJ

grass: 3050 kJ

faeces and urine: 1900 kJ

energy lost $= \dfrac{1900\ kJ}{3050\ kJ} \times 100\% = 62\%$

Figure 14.7 *Only about 4% of the energy in grass eaten by a bullock ends up in new bullock tissues. Most is lost in respiration, faeces and urine.*

The flow of energy through ecosystems

This approach focuses less on individual organisms and food chains and rather more on energy transfer between trophic levels (producers, consumers and decomposers) in the whole ecosystem. There are a number of key ideas that you should understand at the outset:

- Photosynthesis 'fixes' sunlight energy into chemicals such as glucose and starch.

- Respiration releases energy from organic compounds such as glucose.

- Almost all other biological processes (e.g. muscle contraction, growth, reproduction, excretion, active transport) use the energy released in respiration.

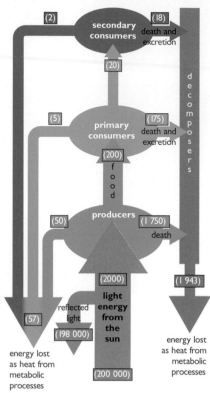

Figure 14.8 *The main ways in which energy is transferred in an ecosystem. The amounts of energy transferred through 1 m² of a grassland ecosystem per year are shown in brackets.*

- If the energy released in respiration is used to produce new cells (general body cells in growth and sex cells in reproduction) then the energy remains 'fixed' in molecules in that organism. It can be passed on to the next trophic level through feeding.

- If the energy released in respiration is used for other processes then it will, once used, eventually escape as heat from the organism. Energy is therefore lost from food chains and webs at each trophic level.

This can be shown in an **energy flow diagram**. Figure 14.8 shows the main ways in which energy is transferred in an ecosystem. It also gives the amounts of energy transferred between the trophic levels of a grassland ecosystem.

As you can see, only about 10% of the energy entering a trophic level is passed on to the next trophic level. This explains why not many food chains have more than five trophic levels. Think of the food chain:

$$A \rightarrow B \rightarrow C \rightarrow D \rightarrow E$$

If we use the idea that only about 10% of the energy entering a trophic level is passed on to the next level, then, of the original 100% reaching A (a producer), 10% passes to B, 1% (10% of 10%) passes to C, 0.1% passes to D and only 0.001% passes to E. There just isn't enough energy left for another trophic level.

Decomposers

The material from dead organisms and the waste products of living organisms must be broken down and recycled. Figure 14.8 shows that this is carried out by decomposers. Decomposers are mainly microorganisms such as fungi and bacteria, and play a key role in recycling. Decomposers break down complex organic materials into simpler substances, which they release into the environment. For example, the respiration of decomposers produces carbon dioxide and water, which can be used by plants for photosynthesis. Fungi and bacteria are also involved in recycling other elements in the ecosystem. Proteins are broken down into ammonia and then converted into nitrates, which are essential mineral ions for plants to use. Humans use these natural recycling processes in the treatment of sewage.

Sewage treatment

Sewage is wet waste from houses, factories and farms. In developed countries where large-scale sewage treatment takes place, industrial and agricultural sewage is usually dealt with separately from household sewage. Household sewage consists of waste water from kitchens and bathrooms and contains human urine and faeces, as well as dissolved organic and inorganic chemicals such as soaps and detergents. It is carried away in pipes called **sewers**, to be treated before it enters waterways such as rivers or the sea.

If sewage is discharged untreated into waterways, it produces two major problems:

- Aerobic bacteria in the water polluted by the sewage use up the dissolved oxygen in the water as they break down the organic materials. This depletion of oxygen kills larger organisms such as freshwater insects and fish.

- Untreated sewage contains pathogenic bacteria, which are a health hazard (see Chapter 13).

When untreated or 'raw' sewage enters a river, the level of oxygen in the water at the sewage outlet becomes very low as the aerobic bacteria and other microorganisms from the sewage decompose the organic matter. Only species that are adapted to live in low-oxygen conditions, such as anaerobic bacteria, can survive. As the water moves away from the outlet, it becomes oxygenated again as it mixes with clean water and absorbs oxygen from the air. The increase in dissolved oxygen levels allows more clean-water species to survive. Figure 14.9 shows the changes in oxygen content, numbers of clean-water animals, and numbers of polluted-water animals downstream from a sewage outlet.

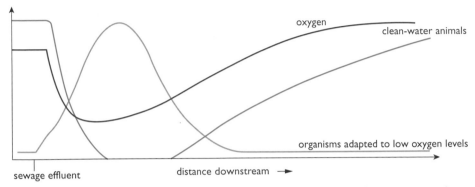

Figure 14.9 *Changes in oxygen levels and types of organism living downstream from a sewage outlet into a river.*

The aim of sewage treatment is to remove solid and suspended organic matter and pathogenic microorganisms, so that a cleaner effluent can be discharged into waterways.

Two methods of sewage treatment are the **percolating (biological) filter method** and the **activated sludge method**. These are compared in Figures 14.10 and 14.11.

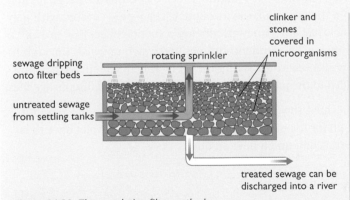

Figure 14.10 *The percolating filter method.*

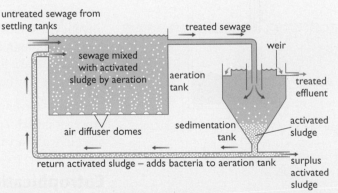

Figure 14.11 *The activated sludge method.*

- Sewage arrives for treatment, is screened to remove large objects, and stands in large settling tanks to allow other solid material to settle out.

- It is then pumped through a pipe rotating over the filter bed, and trickles through the filter. Bacteria, fungi and protozoa in the filter oxidise any organic matter.

- Treated sewage is then discharged into a waterway.

- Sewage arrives for treatment, is screened to remove large objects, and stands in large settling tanks to allow other solid material to settle out.

- It is then passed into an aeration tank, where it is 'activated' by oxygen being pumped in. Bacteria in the tank oxidise the organic material.

- From here it passes to a sedimentation tank, where the activated sludge settles out.

- Some is returned to the aeration tank, and the purified effluent is discharged.

Both methods of treatment rely on a complex ecosystem of aerobic bacteria, fungi, protozoa and other organisms to digest the sewage. Fungi and bacteria start the breakdown process, converting proteins into ammonia and then into nitrates. In turn, protozoa and larger invertebrate animals, such as worms, feed on the bacteria and fungi as well as on the organic matter. The end result is an effluent that contains much less organic matter, and many fewer pathogenic microorganisms. This effluent can be safely discharged into a river. However, it does contain a higher concentration of inorganic ions, such as nitrates and phosphates – and this can produce a different pollution problem: **eutrophication** (see below).

The waste sludge that accumulates in the settling tanks must be treated further to reduce the organic matter and pathogens it contains, before it can be disposed of. There are various ways of achieving this. The commonest method is **anaerobic digestion** by microorganisms in a fermentation tank. This produces **biogas** as a waste product. Biogas is a mixture of methane and carbon dioxide. The methane is produced by anaerobic bacteria called **methanogens**, which means 'methane makers'. Biogas can be used as a fuel in electricity generators or for heating. After anaerobic digestion, the dry, solid material left over can be used for fertiliser or disposed of in landfill sites.

Pit latrines

In areas of the world where there are no sewers or sewage treatment, the **composting toilet** is a simple way of dealing with human waste. There a variety of types, but they all use the same principle. The toilet consists of a hole in the ground over a sunken pit. In the pit, the faeces and urine are broken down by microorganisms. The simplest type of composting toilet is the pit latrine, where the user squats over a shallow trench 1–2 metres deep. An improved version is called the ventilated improved pit latrine (VIP) (Figure 14.12). This has a covered pit and a vent pipe leading from the pit, to take away odours. The pipe is covered with a fly screen. The vent pipe acts like a chimney, producing a suction effect that draws clean air through the latrine. Flies and mosquitoes are drawn upwards through the chimney as well, and prevented from leaving the pit by the fly screen. Earth, sand or sawdust may be added to the pit after use, to prevent smells and discourage flies.

With some larger pit latrines, the large waste pit means that natural decomposition proceeds quickly enough for the pit not to need emptying. Smaller pits have to be filled in with earth periodically, and a new pit dug.

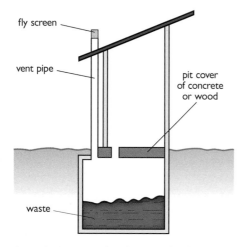

fly screen

vent pipe

pit cover of concrete or wood

waste

Figure 14.12 *A ventilated improved pit latrine.*

Eutrophication

Eutrophication comes from a Greek word meaning 'well fed'. It refers to a situation where large amounts of nutrients enter a body of water, such as a river, lake or even the sea. The nutrients in question are inorganic mineral ions, usually **nitrates** or **phosphates**. Pollution by these minerals can have very harmful effects on an aquatic ecosystem.

There are two main sources of excess minerals:

- from untreated or treated sewage

- from artificial nitrate or phosphate fertilisers.

You have seen how effluent from treated sewage is 'clean', containing little organic material. This is achieved by microorganisms breaking the organic material down into inorganic ions, including nitrate and phosphate – which can lead to eutrophication.

A more serious cause of eutrophication is artificial fertiliser. Streams and rivers that run through agricultural lands that have been treated with fertiliser can contain high concentrations of nitrate and phosphate. This is because nitrate is very soluble in water, and is easily washed out of the soil by rain, a process known as **leaching**. This is less of a problem with phosphate fertiliser, but phosphate is also washed into waterways by surface run-off of water.

The excess mineral ions stimulate the growth of all plants in the river or lake, but this is usually seen first as a rapid growth of algae, called an **algal bloom**. The algae can increase in numbers so rapidly that they form a thick scum on the surface of the water (Figure 14.13).

The algae soon start to die, and are decomposed by aerobic bacteria in the water. Because these bacteria are aerobic, they use up the oxygen in the water. In addition, the algae block the light from reaching other rooted plants, further decreasing the oxygen produced by photosynthesis. The low oxygen levels can result in fish and other aerobic animals dying. In severe cases, the water becomes **anoxic** (containing very little oxygen) and smelly from gases such as hydrogen sulphide and methane from the bacteria. By this stage, only anaerobic bacteria can survive.

A clean water supply

To be safe to drink, water must be clear, low in minerals, and free of harmful chemicals or pathogenic microorganisms. A few sources of water are fit to drink without any treatment, such as water from some mountain springs. But most water needs to go through a treatment process before it is clean enough for use in our homes.

Most human water supplies are taken from rivers or lakes. Sea water is sometimes used as a source when fresh water is not readily available, but first it has to have the salt removed, in a process called desalination.

Whatever the source, water purification treatments are carried out. The process is very similar to sewage treatment, and makes use of the activities of microorganisms (Figure 14.14).

- Water is taken from a source such as a river.

- It is first passed through a **screen**, which is a filter for large solid objects, weeds and other debris.

- Next the water is pumped to a **settling tank**, where large particles in suspension are allowed to settle out. The sludge that settles in the tank has to be removed at intervals and is used for fertiliser, or in landfill. At the sedimentation stage chemicals may be added, such as iron sulfate, to speed up precipitation.

- Next, the key process involving microorganisms takes place. The water is pumped to a **filter bed**, containing graded sizes of particles, from stones and gravels at the bottom, to fine sands at the top of the bed. The water is sprayed onto the filter bed from a revolving arm, and slowly trickles through the graded particles. As it

Figure 14.13 *Algal bloom on a Chinese lake – a major cleanup was needed for the sailing events of the 2008 Beijing Olympics.*

The sequence of events following eutrophication is:

increase in mineral ions

↓

algal bloom

↓

death of algae

↓

decomposition by aerobic bacteria

↓

bacteria use up oxygen

↓

fish and other animals die

does so, bacteria and fungi among the particles break down any organic matter in the water, while protozoa feed on bacteria, including pathogenic species.

- Finally, the water is **chlorinated** (chlorine is added) to kill any remaining microorganisms, before it is stored in covered reservoirs. The covers stop contaminants entering the water (for example, from birds), and also prevent the growth of algae.

- The treated water, now safe for drinking, is then pumped to homes.

The stages in Figure 14.14 are simplified to make them easier to follow – purification may involve several other treatments.

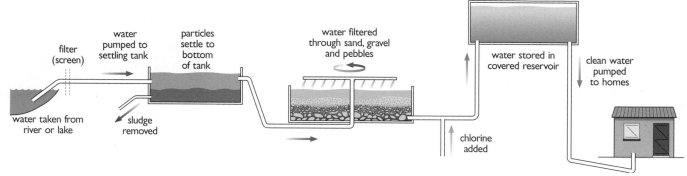

Figure 14.14 *Some of the treatments used to produce clean, drinkable water.*

Pollution

You have seen some examples of water pollution resulting from human activities – the effects of untreated sewage on aquatic ecosystems, and eutrophication caused by fertilisers or treated sewage effluent. These examples illustrate the meaning of pollution:

> *'Pollution takes place when humans release substances into the environment that have harmful effects. The substances are not easily removed by biological processes.'*

A key feature of pollution is the amount of pollutant. Small amounts of pollutants such as carbon dioxide or sulfur dioxide can be absorbed easily and made harmless by the environment. It is the sheer mass of the major pollutants that poses the problem.

Air pollution

We pollute the air with many gases. The main ones are carbon dioxide, carbon monoxide, sulfur dioxide, nitrogen oxides, methane and CFCs (chlorofluorocarbons).

Carbon dioxide

The levels of carbon dioxide have been rising for several hundred years. Over the last 100 years alone, the level of carbon dioxide in the atmosphere has increased by nearly 30%. This recent rise has been due mainly to the increased burning of fossil fuels, such as coal, oil and natural gas, as well as petrol and diesel in vehicle engines. It has been made worse by cutting down large areas of tropical rainforest (see later in this chapter). These extensive forests have been called 'the lungs of the Earth' because they absorb such vast quantities of carbon dioxide and produce equally large amounts of oxygen. Extensive deforestation means that less carbon dioxide is being absorbed. Figure 14.15 shows changes in the level of carbon dioxide in the atmosphere (in parts per million) from 1960 to 2010.

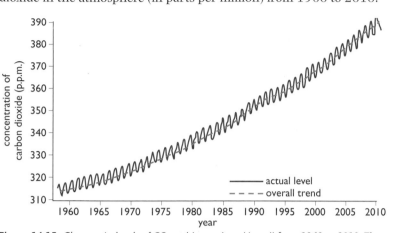

Figure 14.15 *Changes in levels of CO$_2$ at Mauna Loa, Hawaii from 1960 to 2010. The peaks and troughs along the curve are explained in Figure 14.16. The data are from the US National Oceanic and Atmospheric Administration.*

The increased levels of carbon dioxide and other gases contribute to **global warming**, or the **enhanced greenhouse effect**. It is important to understand that the 'normal' greenhouse effect is a natural phenomenon – without it, the surface temperature of the Earth would be about 30 °C lower than it is, and life as we know it would be impossible.

Carbon dioxide is one of the 'greenhouse gases' that are present in the Earth's upper atmosphere. Other greenhouse gases include water vapour (H$_2$O), methane (CH$_4$), nitrous oxide (N$_2$O) and chlorofluorocarbons (CFCs). Most greenhouse gases occur naturally, while some (like CFCs) are produced only by human activities.

The 'natural' greenhouse effect is shown in Figure 14.17.

In any one year, there is a peak and a trough in the levels of carbon dioxide in the atmosphere. This is shown more clearly in Figure 14.16.

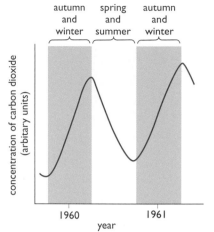

Figure 14.16 *Seasonal fluctuations in carbon dioxide levels.*

In the autumn and winter, trees lose their leaves. They photosynthesise much less and so absorb little carbon dioxide. They still respire, which produces carbon dioxide, so in the winter months, they give out carbon dioxide and the level in the atmosphere rises. In the spring and summer, with new leaves, the trees photosynthesise faster than they respire. As a result, they absorb carbon dioxide from the atmosphere and the level decreases. However, because there are fewer trees overall, it doesn't quite get back to the low level of the previous summer.

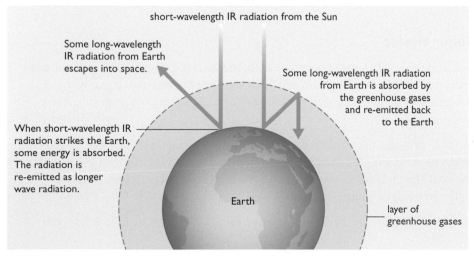

Figure 14.17 *The greenhouse effect.*

Short-wavelength infrared (IR) radiation from the Sun reaches the Earth. Some is absorbed and some is re-emitted as longer-wavelength IR radiation. The greenhouse gases absorb and then re-emit some of this long-wavelength IR radiation, which would otherwise escape into space. This then heats up the surface of the Earth.

Infrared radiation is radiated heat. The heat you can feel from an electric fire is IR radiation.

The problem is that human activities are polluting the atmosphere with extra greenhouse gases such as carbon dioxide. This is thought to be causing a rise in the Earth's surface temperature, hence the *enhanced* greenhouse effect, or global warming.

A rise in the Earth's temperature of only a few degrees would have many effects:

- Polar ice caps would melt and sea levels would rise.

- A change in the major ocean currents would result in warm water being redirected into previously cooler areas.

- A change in global rainfall patterns could result. With all the extra water in the seas, there would be more evaporation from the surface and so more rainfall in most areas. Long-term **climate change** could occur.

- It could change the nature of many ecosystems. If species could not migrate quickly enough to a new, appropriate habitat, or adapt quickly enough to the changed conditions in their current habitat, they could become extinct.

- Changes in agricultural practices would be necessary as some pests became more abundant. Higher temperatures might allow some pests to complete their life cycles more quickly.

Carbon monoxide

In many countries there are now strict laws controlling the permitted levels of carbon monoxide in the exhaust gases produced by newly designed engines.

When substances containing carbon are burned in a limited supply of oxygen, carbon monoxide (CO) is formed. This happens when petrol and diesel are burned in vehicle engines. Exhaust gases contain significant amounts of carbon monoxide. It is a dangerous pollutant as it is colourless, odourless and tasteless and can cause death by asphyxiation. Haemoglobin binds more strongly with carbon monoxide than with oxygen. If a person inhales carbon monoxide for a period of time, more and more haemoglobin becomes bound to carbon monoxide and so cannot bind with oxygen. The person may lose consciousness and, eventually, may die as a result of a lack of oxygen.

Sulfur dioxide

Sulfur dioxide (SO_2) is an important pollutant as it is a major constituent of **acid rain** (see below). It is formed when fossil fuels are burned, and it can be carried hundreds of miles in the atmosphere before finally combining with rainwater to form acid rain.

Some lichens are more tolerant of sulfur dioxide than others. In some countries, patterns of lichen growth can be used to monitor the level of pollution by sulfur dioxide. The different lichens are called **indicator species** as they 'indicate' different levels of sulfur dioxide pollution (see Figure 14.18).

The map in Figure 14.18 shows zones in Britain colonised by different types of lichen.

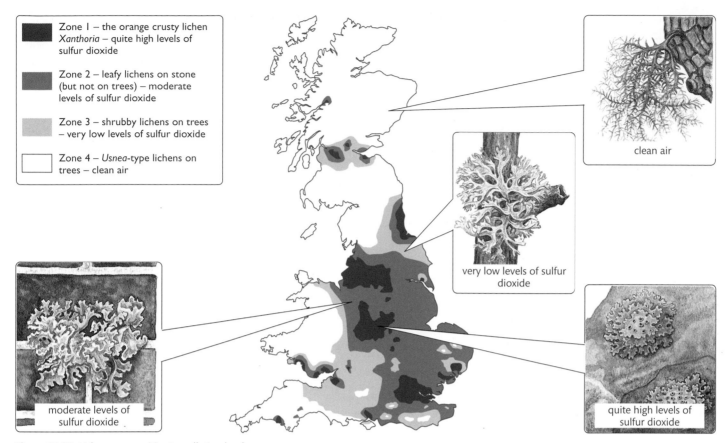

Figure 14.18 *Lichens are sensitive to pollution levels.*

Nitrogen oxides

Nitrogen oxides (NO_x) are also constituents of acid rain. They are formed when petrol and diesel are burned in vehicle engines.

Methane

Methane (CH_4) is an organic gas. It is produced when microorganisms ferment larger organic molecules to release energy. The most significant origins of these microorganisms are:

Herds of cattle can produce up to 40 dm³ of methane per animal per hour.

- decomposition of waste in landfill sites by microorganisms

- fermentation by microorganisms in the rumen of cattle and other ruminants

- fermentation by bacteria in rice paddy fields.

Methane is a greenhouse gas, with effects similar to carbon dioxide. Although there is less methane in the atmosphere than carbon dioxide, each molecule has a bigger greenhouse effect.

Besides its effects on living things, acid rain causes damage to stonework by reacting with carbonates and other ions in the stone (Figure 14.20).

Figure 14.20 *The features of this stone lion have been dissolved by acid rain.*

Acid rain

Rain normally has a pH of about 5.5 – it is slightly acidic due to the carbon dioxide dissolved in it. Both sulfur dioxide and nitrogen oxides dissolve in rainwater to form a mixture of acids, including sulfuric acid and nitric acid. As a result, the rainwater is more acidic with a much lower pH than normal rain (Figure 14.19).

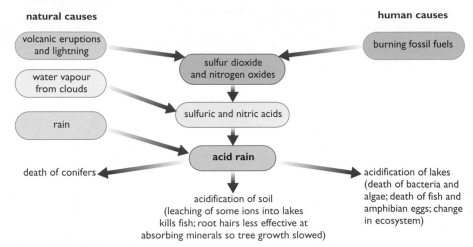

Figure 14.19 *The formation of acid rain and its effects on living organisms.*

Deforestation

The last great natural forests of the world are the **tropical rainforests**. These form a belt around the equatorial regions of the Earth, in South America, central Africa and Indonesia. These rainforests are rapidly being destroyed by humans, in a process called **deforestation**. Deforestation is a consequence of the enormous growth of the human population. Every year, tens of thousands of hectares of rainforest are cut down to provide wood or to clear the land for farming (Figure 14.21). Much of the clearing is done by 'slash-and-burn' methods, adding to the carbon dioxide in the atmosphere and contributing to global warming. It also removes the trees, which would otherwise be absorbing carbon dioxide for photosynthesis. Deforestation is adding to global warming and climate change.

Apart from adding to global warming, there are a wide range of other problems caused by deforestation – some of these are listed below.

- Destruction of habitats and reduced biodiversity. Rainforests are home to millions of species of plants, animals and other organisms. It has been estimated that 50–70% of species living on the Earth inhabit rainforests.

- Reduced soil quality. There are no trees and other plants to return minerals to the soil when they die, and no tree roots to hold the soil together. Crops planted in deforested areas rapidly use up minerals from the soil, and rain washes the minerals out (leaching).

Figure 14.21 *Deforestation.*

- The soil is exposed due to lack of a canopy, and is blown or washed away (Figure 14.22). Soil may be washed into rivers, causing rising water levels and flooding of lowland areas.

- Deforestation may produce climate change. Trees are an important part of the Earth's water cycle, returning water vapour from the soil to the air through their leaves. Cutting down the forests will upset the water cycle.

- In the past, rainforests have been a valuable source of many medicinal drugs, as well as species of plant that have been cultivated as crops. There are probably many undiscovered drugs and crop plants that will be lost with the deforestation.

You can see that there are many reasons why conservation of the remaining rainforests is urgently needed. Sometimes controlled replanting schemes (**reforestation** or **re-afforestation**) are carried out, allowing for sustainable timber production.

Tackling the problem of clearance for farming is a more complex issue. Large-scale cattle ranching on deforested land is carried out mainly to supply meat for the burger industry, or palm oil for cosmetics, and you may feel these reasons are unethical. The small-scale farming by poor farmers around the edges of rainforests is more understandable. These farmers are poor, and their livelihood and that of their family depends on this way of life. The only alternative would be to resettle the farmers and give them financial help to establish farms in other areas.

Cutting down and burning rainforests is adding to carbon dioxide in the atmosphere. But does it affect oxygen? Books often talk about the rainforests as being the 'lungs of the world' because of the large amount of oxygen they produce. But scientists think that overall they don't have much effect on the world's oxygen levels. In a rainforest, the oxygen produced by the living plants is roughly balanced by the oxygen consumed by decomposers feeding on dead plant material.

Figure 14.22 *This river in Madagascar is full of silt from erosion caused by deforestation.*

End of Chapter Checklist

You should now be able to:

✓ understand the relationship between humans and their environment

✓ explain the dependence on green plants for supplies of food and oxygen

✓ recall the word equation to summarise the process of photosynthesis

✓ recall the names given to different trophic levels, including producers, primary, secondary and tertiary consumers, and decomposers

✓ describe the transfer of substances and of energy along a food chain

✓ understand that only about 10% of energy is transferred from one trophic level to the next

✓ understand the biological consequences of pollution of water by untreated sewage, including increases in the number of microorganisms causing depletion of oxygen

✓ understand the role of non-pathogenic bacteria and fungi in the decomposition of organic matter

✓ recall the processes of sewage treatment in a modern sewage works and in a pit latrine

✓ explain the role of aerobic and anaerobic microorganisms in sewage breakdown

✓ explain how eutrophication can result from leached minerals from excess nitrogen fertiliser or treated sewage

✓ understand the scientific principles used in the purification, distribution and storage of water

✓ understand the consequences of pollution of air by sulfur dioxide and carbon monoxide

✓ recall that carbon dioxide, water vapour, nitrous oxide, methane and CFCs are greenhouse gases

✓ explain how human activities contribute to greenhouse gases

✓ explain that an increase in greenhouse gases results in an enhanced greenhouse effect, and that this may lead to global warming and its consequences

✓ understand the effects of deforestation, including leaching, soil erosion, disturbance of the water cycle, and the balance of atmospheric oxygen and carbon dioxide

Questions

1 A marine food chain is shown below (copepods are small, shrimp-like crustaceans).

algae → copepods → herring → salmon → human

a) Which organism is:

  i)   the producer

  ii) the secondary consumer?

b) What is the importance of the producer in any food chain?

c) Most food chains consist of organisms at only three or four trophic levels. Suggest why there are five in this example.

2 In a year, 1 m² of grass produces 21 500 kJ of energy. The diagram below shows the fate of the energy transferred to a cow feeding on the grass.

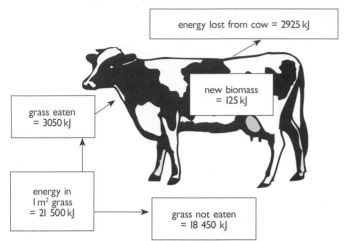

a) Calculate the energy efficiency of the cow from the following equation.

$$\text{energy efficiency} = \frac{\text{energy that ends up as part of cow's biomass}}{\text{energy available}} \times 100$$

b) State two ways in which energy is lost from the cow.

c) Suggest what may happen to the 18 450 kJ of energy in the grass that was not eaten by the cow.

3 Read the following description of the ecosystem of a mangrove swamp.

Pieces of dead leaves (detritus) from mangrove plants in the water are fed on by a range of crabs, shrimps and worms. These, in turn, are fed on by young butterfly fish, angelfish, tarpon, snappers and barracuda. Mature snappers and tarpon are caught by fishermen as the fish move out from the swamps to the open seas.

a) Use the description to construct a food web of the mangrove swamp ecosystem.

b) Write out two food chains, each containing four organisms from this food web. Label each organism in each food chain as producer, primary consumer, secondary consumer or tertiary consumer.

c) Decomposers make carbon in the detritus available again to mangrove plants.

i) In what form is this carbon made available to the mangrove plants?

ii) Explain how the decomposers make the carbon available.

4 Some untreated sewage is accidentally discharged into a small river. A short time afterwards, a number of dead fish are seen at the spot. Explain, as fully as you can, how the discharge could lead to the death of the fish.

5 The flowchart below shows stages in the percolated filter method of sewage treatment.

untreated sewage
↓
screening
↓
settling tank
↓
filter bed
↓
treated effluent discharged into river

a) What is the function of the settling tank?

b) Sludge from the settling tank can be treated by anaerobic fermentation. Name two useful products of this process.

c) Why is it important that conditions are aerobic in the filter bed?

d) Name one pollutant still present in the treated sewage effluent that could cause eutrophication in the river.

6 The diagram shows the profile of the ground on a farm either side of a pond.

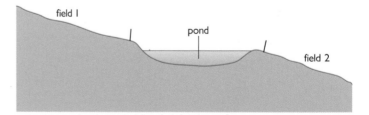

The farmer applies nitrate fertiliser to the two fields in alternate years. When he applies the fertiliser to field 1, the pond often develops an algal bloom. This does not happen when fertiliser is applied to field 2.

a) Explain why an algal bloom develops when he applies the fertiliser to field 1.

b) Explain why no algal bloom develops when he applies the fertiliser to field 2.

c) Explain why the algal bloom is more pronounced in hot weather.

7 The graph shows the changing concentrations of carbon dioxide at Mauna Loa, Hawaii over a number of years.

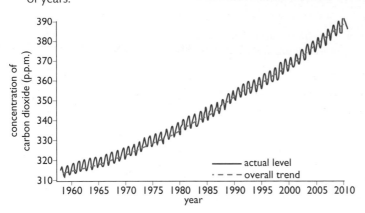

a) Describe the overall trend shown by the graph.

b) Explain the trend described in *a)*.

c) In any one year, the level of atmospheric carbon dioxide shows a peak and a trough. Explain why.

8 The diagram shows how the greenhouse effect is thought to operate.

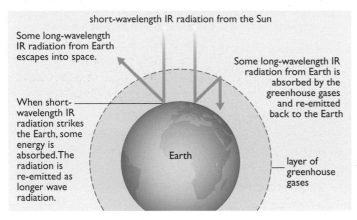

a) Name two greenhouse gases.

b) Explain one benefit to the Earth of the greenhouse effect.

c) Suggest why global warming may lead to malaria becoming more common in Europe.

9 The table gives information about the pollutants produced in extracting aluminium from its ore (bauxite) and in recycling aluminium.

Pollutants	Amount (g per kg aluminium produced)	
	Extraction from bauxite	Recycling aluminium
Air:		
sulfur dioxide	88 600	886
nitrogen oxides	139 000	6 760
carbon monoxide	34 600	2 440
Water:		
dissolved solids	18 600	575
suspended solids	1 600	175

a) Calculate the percentage reduction in sulfur dioxide pollution by recycling aluminium.

b) Explain how extraction of aluminium from bauxite may contribute to the acidification of water hundreds of miles from the extraction plant.

c) Suggest two reasons why there may be little plant life in water near an extraction.

Appendix A: A Guide to Exam Questions on Practical Work

Why is this appendix important?

This appendix is designed to help you gain the 20% of marks allotted to the questions on practical work. The practicals that you should know about are set out in the specification and are fully described in this book, along with some extra ones. In the exams, you may be asked questions based on these, although the apparatus shown may be slightly different. However, you will also be given questions that are about unfamiliar biological investigations. Don't be worried by these – you are not expected to have carried them out, just to be able to apply your knowledge and understanding to interpret experimental design or analyse data.

Questions in the exams will generally cover the following areas:

- understanding safety precautions
- recognising apparatus and understanding how to use it
- recalling tests for certain substances
- manipulating and analysing data
- plotting and interpreting graphs
- understanding experimental design
- planning experiments.

Safety precautions

As part of a question, you may be asked to comment on appropriate safety precautions. The most obvious precaution is to **wear eye protection**. Eye protection must be worn whenever chemicals or Bunsen burners are used. This will apply to many experiments or investigations, with a few exceptions such as measuring heart rate. Some other examples of safety precautions, and the reason they are taken, are shown in the table.

Precaution	Reason
wash hands after handling biological material such as enzymes	to avoid contamination
keep flammable liquids such as ethanol away from a naked flame	to avoid the ethanol catching fire
take care with fragile glassware such as pipettes, microscope cover slips, etc.	to avoid cutting yourself
do not touch electrical apparatus (such as a microscope with built-in lamp) with wet hands	to avoid getting an electric shock
use a water bath to heat a test tube of water, rather than heating it directly in a Bunsen flame	to avoid the heated liquid jumping out of the tube

This book is written for the new Edexcel IGCSE Human Biology specification, which will be examined for the first time in June 2011. At first there will be a shortage of past papers covering the new specification, so you will have to look at similar questions from past papers for previous Human Biology (or Biology) specifications.

In the new specification, there is no separate paper examining investigative skills. Questions on practical work are mixed up with theory questions in both paper 1 and paper 2.

Specimen exam papers 1 and 2 are provided on Edexcel's subject website for IGCSE Human Biology. The website will also provide advice about how to get hold of exam papers after June 2011, as soon as they become available.

Recognising apparatus and demonstrating understanding of how to use it

One of the simplest types of question in the exams will require you to recognise common pieces of laboratory apparatus, such as a Bunsen burner, thermometer, measuring cylinder or stopwatch. You will also need to recognise particular 'biological' apparatus such as a microscope, bench lamp and so on, and how they are used. If you have been able to carry out most of the practical work in the specification, this kind of question is very straightforward, for example:

a) What is the name of this piece of apparatus? *(1 mark)*

b) Draw a line on the apparatus to show a volume of 30 cm³ *(1 mark)*

Note: Use a pencil to draw any lines on diagrams, graphs, etc. Then you can rub it out if you make a mistake. On this measuring cylinder, a horizontal line should be drawn halfway between the 20 and 40 cm³ – it's that easy!

Recalling tests for substances

There are only five chemical tests in the specification. These are for:

- starch – using iodine solution

- glucose (or 'reducing sugar') – using Benedict's solution

- protein – using biuret solutions

- lipid – using ethanol

- carbon dioxide – using hydrogencarbonate indicator or limewater.

The tests for starch, glucose, protein and lipid are described on pages 38–39.

The effects of carbon dioxide on hydrogencarbonate indicator and limewater are described on page 24.

Once you have learnt these, the questions are easy! For example:

John decided to test some milk for glucose.

a) Describe the test he would do. *(2 marks)*

b) What result would he see if glucose was present? *(1 mark)*

c) Suggest how he might use the results to say how much glucose was present. *(1 mark)*

Note: You would get one mark for 'Use Benedict's solution'; one for 'heating'; and one for the colour produced (orange or similar). In (c) the depth of colour would show how much glucose was present.

Manipulating and analysing data

Some questions in an exam will involve analysing data. This means you will be provided with some results from an experiment, and be asked to process it in various ways, such as:

- putting raw data from a notebook into an ordered table
- counting numbers of observations
- summing totals
- calculating an average
- calculating a missing value
- identifying anomalous results.

For example:

Kirsty monitored how her temperature changed during exercise. She took her temperature before she started the exercise, and every 2 minutes during the exercise. Here are her results:

Before exercise 36.4 °C   At 12 minutes = 37.6

37.3 after 10 mins, 2 mins = 36.8

After 4 mins = 37.1, 6 min = 37.2, 8 min = 36.9 °C

a) Organise Kirsty's results into a table.   *(4 marks)*

b) Identify the time when an anomalous temperature reading was taken.   *(1 mark)*

Marks for the answer are awarded like this:

Table with time and temperature headings *(1 mark)*

a)

Time (min)	Temperature (°C)
0	36.4
2	36.8
4	37.1
6	37.2
8	36.9
10	37.3
12	37.6

Units in header row *(1 mark)*

Readings in order *(1 mark)*

Two columns *(1 mark)*

b) The anomalous temperature reading was taken at 8 minutes. 'Anomalous' means that the result doesn't fit into the pattern of the other results. In this case, the temperature reading at 8 minutes is lower than expected. You might have to circle an anomalous result in a table or on a graph – see below.

You may have to calculate totals, averages, percentages or missing values from a table. For example, the table below contains results from an investigation into diffusion. The table shows the time taken for different cubes of agar jelly containing potassium permanganate to turn colourless when placed in beakers of dilute hydrochloric acid. Three groups of students carried out the same experiment. The table is incomplete.

Cube size (side length, cm)	Time taken for cube to turn colourless (minutes)				Mean time (min)
	Group 1	Group 2	Group 3	Total for three groups	
2	52	45	49	?	?
1	10	15	13	38	12.7
0.5	6	5	?	19	6.3

a) For the 2 cm cubes, calculate the missing value in the 'Total' column. *(1 mark)*

b) For the 2 cm cubes, calculate the mean time for the three groups. *(1 mark)*

c) For the 0.5 cm cubes, calculate the missing value for Group 3. *(1 mark)*

The answers are:

a) Total = 52 + 45 + 49 = 146 minutes

b) Mean time = total/3 = 48.7 minutes

(Use the same number of decimal places as in the given answers.)

c) Missing value = 19 − (6 + 5) = 8 minutes

This is straightforward arithmetic: just be careful!

If you have to find a percentage change, this is calculated from:

$$\frac{(\text{final value} - \text{starting value})}{\text{starting value}} \times 100$$

For example:

After exercise, a boy's heart rate increased from 75 beats per minute to 97 beats per minute. What is the percentage increase in his heart rate?

The answer is:

$$\% \text{ increase} = \frac{(97 - 75)}{75} \times 100 = 29.3\%$$

Plotting and interpreting graphs

In biology (unlike in chemistry or physics), a line graph is usually constructed by joining the points with straight lines. It is rare to have data that produce an overall straight line, and 'best fit' lines are not usually appropriate either.

Take the following example. The table below shows the activity of the enzyme amylase at different temperatures. The activity is measured as cm^3 of starch suspension digested per minute.

Plot a graph of these results.

Temperature (°C)	Activity of amylase (cm³ per min)
10	0.14
20	0.53
30	0.71
40	5.00
50	1.43
60	0.59
70	0.23

A graph of these data looks like this:

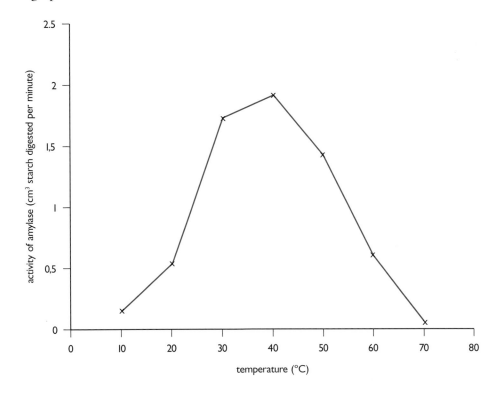

It is important to use a ruler and a sharp pencil for drawing graphs. A pen is no good, and if you make a mistake with a pen you can't rub it out. Remember to label the axes and include units.

The variable that is set by the person designing the experiment is temperature. This is called the **independent** variable, and goes on the horizontal or x-axis. The **dependent** variable is activity of the enzyme: this goes on the vertical or y-axis.

You will probably have to describe the pattern shown by the graph, or identify a point on the graph where something happens. For instance in the graph above, the rate of reaction (activity of the enzyme) increases with temperature from 10 to 40 °C. Above 40 °C, the rate decreases again until 70 °C, where the reaction has almost stopped. The optimum temperature for this enzyme is about 40 °C.

A question like this would go on to ask about the biology behind the graph – this one's all about the effect of temperature on enzyme activity (see page 4).

Note that in many cases with 'biological' graphs, the line doesn't pass through the 0,0 coordinates.

You may also have to draw a bar chart from data you are given.

Experimental design

Questions may test various aspects of experimental design, especially:

- the meaning of a 'fair test' and controlled variables

- how to improve accuracy

- how to improve reliability.

A **fair test** is an experiment in which only one thing is changed at a time. For example, imagine you are carrying out an investigation to compare the energy content of three different foods (pasta, biscuit and chocolate) using the apparatus on the left.

The factor you are changing is the type of food. It is important that all other factors are kept constant. For example, the following should all be the same for each food tested:

- volume of water in the test tube

- starting temperature of the water

- type of test tube and thermometer

- distance of the burning food from the tube

- the way the food is ignited.

Ideally, the mass of each food should be the same, although it is easier to weigh each sample and correct for any difference in mass by calculating the energy content per gram of food.

The factors above are called **controlled variables**, because you, the experimenter, are controlling them (keeping them from changing) so that they shouldn't affect the results.

Some biological experiments involve the use of a **control**. This is an experimental set-up where the key factor is missing. For example, if you were investigating the effect of the enzyme amylase on starch (as in 'Plotting and interpreting graphs', above), the experiment would involve mixing amylase with starch to see if the amylase breaks the starch down. It is possible (but unlikely) that the starch might break down on its own, so a control would be a tube with just starch, to check whether this happens. A better control would be starch with *boiled* amylase, which is as near as possible to the experimental set-up, but without the amylase being able to act as an enzyme.

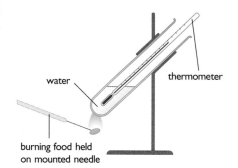

water

thermometer

burning food held
on mounted needle

Hopefully, findings from experiments should be both **accurate** and **reliable**. Students often confuse these two terms.

Accuracy refers to how 'correct' the results are. It depends mainly on the methods you use to obtain them.

For example, the starch–amylase experiment measures amylase activity by how quickly it breaks down the starch, as shown by a colour change with iodine solution (see page 5). As the starch is broken down, the colour of the iodine changes from blue-black to brown, and finally to yellow. This is a gradual process, and the experimenter has to judge the colour change by eye. It may not be possible to do this better than to the nearest half a minute, as in the table on page 6). So the accuracy is limited by the method used. (It is possible to measure the colour change in a piece of apparatus called a colorimeter – this would increase the accuracy, but is beyond what you could be expected to do for an experiment at IGCSE level.)

Sometimes you have to accept that accuracy will be limited by the apparatus available. But in the exam, you might need to comment on how accuracy could be improved, for example by using more sensitive measuring equipment.

Reliability is a measure of how similar the results are, if you carried out the same experiment several times. It tells you how confident you can be that your results are correct. For example, imagine you measured the activity of an enzyme at $40\,°C$ eight times, using two different sets of apparatus, and obtained these results:

Measurement	Activity (arbitrary units)*	
	Apparatus 1	**Apparatus 2**
1	26	14
2	25	19
3	27	52
4	22	44
5	28	12
6	25	10
7	26	37
8	26	21
Mean	25.6	26.1

*'Arbitrary units' means that the units here are not important – but they are the same in both columns, allowing comparisons to be made.

Which mean value in the table is a more reliable measure of the activity? You can tell that the mean for Apparatus 1 is more reliable, because the individual readings are closer together. The only way you can tell if results are reliable is to carry out repeats. You should make sure you understand the difference between accuracy and reliability.

Planning an experiment

You may have to give a short account of how you would carry out an experiment. This might be based on an experiment in the book, or it might involve an unfamiliar situation. Take this example:

Describe an experiment you could carry out to find out the effect of changing the concentration of the enzyme amylase on the rate of starch digestion.   (6 marks)

This is similar to the experiment described on page 5 in this book, but involves changing the *concentration* of amylase, rather than the *temperature*. In any question like this, you have to answer the following questions.

- What are you going to change?

- What are you going to keep constant (control) during the experiment?

- What are you going to measure, and how?

- How are you going to check that the results are reliable?

A really good answer would be:

I would make up five different concentrations of amylase solution (e.g. 1%, 2%, 3%, 4% and 5%) using the same source of enzyme, such as fungal amylase. I would mix 10 cm³ of each concentration with 10 cm³ of 1% starch suspension, and time how long it takes for the amylase to digest the starch. To tell when the starch has been digested, I would add one drop of iodine solution to the mixture. The iodine would start off blue-black when the starch is present, but change colour to yellow when the starch has been digested, so I would time how long it takes for this to happen. I would use a water bath to keep the temperature of the reaction mixture at 40 °C. To ensure reliability, I would repeat the experiment three times at each concentration and compare the results.

The marks (up to a maximum of 6) could be awarded for:

- using two or more concentrations of amylase

- using equal volumes of amylase

- using the same source of amylase, or explaining its source

- using the same concentration or volume of starch

- keeping the temperature constant

- using iodine solution to show a colour change (an alternative would be to use Benedict's solution to test for the formation of reducing sugar)

- change from blue-black to yellow

- measuring the time taken for the colour change

- repeating tests at each concentration (for reliability).

Appendix B: Exam Tips

Where to revise
Somewhere quiet away from distractions. You need good lighting and you should be comfortable.

When to revise
As soon as possible after school finishes each day, before you get too tired to work effectively.

How to revise
Have a plan and stick to it. Think about how many topics you need to cover. Make a list of all the topics and mark those that you find difficult. Revise a mixture of easy and more challenging topics each day. This will give you the satisfaction of making progress every day. When you have revised something, tick it on your list. This will show clearly how much still needs to be done.

Divide your time
Working for a whole evening without a break will achieve less than dividing your time into segments. Split your evening into revision time and leisure breaks. Take at least 10 minutes off for each hour worked. When you start again, take 5 minutes to review the previous topic, then move on.

Group revision
Some students find that revision is less boring and more effective in a group. Ask each other questions, choose topics in turn, share good ideas.

Organising the information
Think of ways to organise the information you need to revise. You can use summaries, checklists, file cards, sticky-notes, flow-schemes and key words.

Past paper questions
Read examination papers from previous years. This will show you the styles of questions and indicate which topics are more likely to be repeated. Look carefully at the marks printed on the paper and the spaces left for answers. Fill the space provided and make points according to marks available. For example, make two points if there are two marks available.

Mathematical requirements
You will be awarded marks for every part of the answer that is correct, even if the final answer is wrong. Always have a go and show your working. Don't forget to state the units you are working in.

> *Very few people enjoy revising for examinations. Being very well organised with your revision will mean you revise for as little time as possible. Revising for hours on end without a break is not productive – your concentration will not last and you will not retain the facts.*

Appendix C: More Practice Questions

Part 1: Multiple choice questions

1 Which of the following is present in the nucleus of a cell, and is made up of genes?

   A DNA

   B mitochondrion

   C chromosome

   D allele

2 Which of the following is a tissue?

   A pancreas

   B skin

   C testis

   D blood

3 'Movement of molecules against a concentration gradient, using energy from respiration' is called:

   A osmosis.

   B active transport.

   C diffusion.

   D mass flow.

4 Which of the following statements about aerobic respiration are true? Aerobic respiration produces:

   I oxygen.

   II carbon dioxide.

   III energy.

   IV lactic acid.

   A I, II, III and IV

   B I, II and III

   C II and III

   D II, III and IV

5 Which of the following changes take place during inhalation (breathing in)?

   A The diaphragm contracts, the internal intercostals contract, the pressure in the chest cavity decreases.

   B The diaphragm relaxes, the external intercostals contract, the pressure in the chest cavity decreases.

   C The diaphragm relaxes, the internal intercostals relax, the pressure in the chest cavity increases.

   D The diaphragm contracts, the external intercostals contract, the pressure in the chest cavity decreases.

6 Which of the following is **not** a feature of alveoli?

   A large surface area

   B close proximity to blood vessels

   C thick walls

   D moist lining

7 The diagram shows a section through the heart.

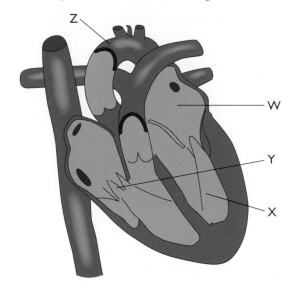

The structures labelled W, X, Y and Z are:

   A W – left atrium, X – left ventricle, Y – tricuspid valve, Z – aorta.

   B W – right atrium, X – left ventricle, Y – tricuspid valve, Z – pulmonary artery.

   C W – left atrium, X – bicuspid valve, Y – tricuspid valve, Z – aorta.

   D W – left ventricle, X – left atrium, Y – bicuspid valve, Z – vena cava.

8 A girl is sitting under a shady tree, reading a book. She looks up into the sunny sky at an aeroplane. Which of the following changes will take place in her eyes?

   A The pupils constrict and the lens becomes less convex.

   B The pupils dilate and the lens becomes more convex.

   C The pupils dilate and the lens becomes less convex.

   D The pupils constrict and the lens becomes more convex.

9 During the knee-jerk reflex, which is the following is the correct order of events?

A effector → motor neurone → CNS → sensory neurone → receptor

B effector → sensory neurone → CNS → motor neurone → receptor

C receptor → motor neurone → CNS → sensory neurone → effector

D receptor → sensory neurone → CNS → motor neurone → effector

10 Which of the following will **not** happen when the hormone adrenaline is released?

A increase in heart rate

B increase in blood flow to the gut

C dilation of the pupils

D increase in breathing rate

11 Muscles move bones by:

A relaxing and pulling.

B relaxing and pushing.

C contracting and pushing.

D contracting and pulling.

12 Which of the following happens when the human body temperature rises above normal?

A Blood vessels in the skin constrict and blood flow decreases.

B Blood vessels in the skin constrict and blood flow increases.

C Blood vessels in the skin dilate and blood flow decreases.

D Blood vessels in the skin dilate and blood flow increases.

13 Which of the following statements about the action of antidiuretic hormone are true?

I More ADH is released when the water content of the blood rises.

II ADH increases the permeability of the collecting duct.

III When more ADH is released, more water is reabsorbed.

A I and II

B I and III

C II and III

D I, II and III

14 In the diagram of a section through the kidney, which part contains the glomeruli?

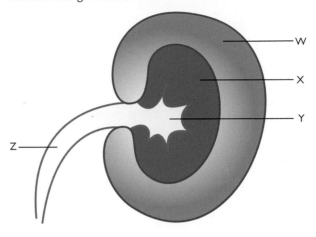

A W

B X

C Y

D Z

15 Which of the following is **not** a function of the ovaries?

A production of follicle stimulating hormone (FSH)

B production of oestrogen

C production of progesterone

D production of egg cells

16 Sex cells are:

A diploid cells produced by meiosis.

B diploid cells produced by mitosis.

C haploid cells produced by meiosis.

D haploid cells produced by mitosis.

17 Alleles **B** and **b** are codominant. Two heterozygous individuals are crossed. What will be the ratio of phenotypes in the offspring?

A 3 : 1

B 1 : 2 : 1

C 1 : 1 : 1 : 1

D 1 : 1

18 Malaria is caused by:

A mosquitoes.

B bacteria.

C a virus.

D a protozoan.

19 AIDS is nearly always fatal because:

   A it damages the body's immune system.

   B it causes cancers to develop.

   C it infects all the major organ systems.

   D it causes heart failure.

20 The greenhouse effect is caused by:

   A greenhouse gases reflecting long wavelength infrared radiation.

   B greenhouse gases emitting long wavelength infrared radiation.

   C greenhouse gases emitting short wavelength infrared radiation.

   D greenhouse gases reflecting short wavelength infrared radiation.

Part 2: Longer structured questions requiring information from more than one chapter

1 Copy and complete the following account:

One way in which plant cells differ from animal cells is that plant cells have cell walls made of _____. In food chains, plants are known as producers because they can carry out the process of _____, which converts carbon dioxide and water into organic molecules. The first product is the sugar _____, which can be respired for energy. The cells of both animals and plants have nuclei containing chromosomes. Chromosomes contain the genetic material called _____, which codes for the production of _____ in a cell. Bacteria lack a true nucleus, having their genetic material in a single chromosome that is loose in the cytoplasm. They also carry genes in small rings called _____. Many bacteria and fungi break down organic matter in the soil. They are known as _____. Some bacteria are pathogens, which means that they _____.

2 Digestion is brought about by enzymes converting large, insoluble molecules into smaller molecules that can be more easily absorbed into the blood.

   a) The activity of enzymes is affected by pH and temperature. The graph shows the activity of two human enzymes from different regions of the gut at different pHs.

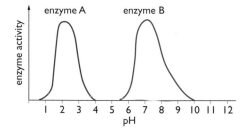

   i) Suggest which regions of the gut the two enzymes come from. Explain your answer.

   ii) Which nutrient does enzyme A digest?

   b) Farmers sometimes include **urea** in cattle food. This increases their growth rate, providing more food for humans. Microorganisms in the stomachs of the cattle can use urea to make protein.

   i) Suggest how feeding urea to cattle can increase their rate of growth.

   ii) Which organ in the human body makes urea?

   iii) How is urea made in this organ?

   iv) The Bowman's capsule and the Loop of Henlé are two parts of a kidney nephron. Explain how each of them helps to remove urea from the blood.

   v) Explain why removal of urea in the urine is an example of excretion.

3 The graph shows the changes in growth rate in humans from birth to adulthood.

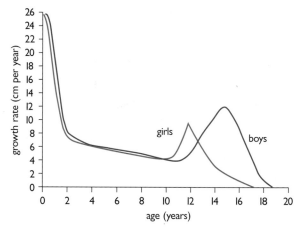

   a) From the graph, when is the growth rate the greatest?

   b) Describe and explain the changes in growth that occur between the ages of 10 and 18 in:

   i) girls

   ii) boys.

   c) State two ways in which growth in humans can be measured. Suggest one advantage and one disadvantage of each method.

   d) Different parts of the body grow at different rates. Suggest reasons why:

   i) lymph tissue (which includes tissues involved in immune responses) grows quicker during childhood than many other tissues

   ii) the head and brain are almost full size by the age of 10 years.

   e) What is meant by the 'secondary sexual characteristics'? Explain how the development of secondary sexual characteristics is controlled in boys and girls.

4 Fish called herring form an important food supply for humans. The diagram shows a simplified food web of the adult herring.

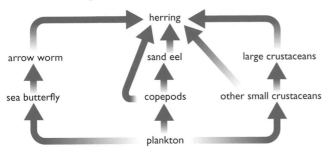

a) i) From the above food web, write out a food chain containing four organisms.

ii) From your food chain, name the primary consumer and secondary consumer.

iii) Name one organism in the web that is both a secondary consumer and a tertiary consumer. Explain your answer.

b) The amount of energy in each trophic level has been provided for the following food chain. The units are kJ/m²/year.

plankton (8869) → copepod (892) → herring (91)

i) Sketch a pyramid of energy for this food chain.

ii) Calculate the percentage of energy entering the plankton that passes to the copepod.

iii) Calculate the percentage of energy entering the copepod that passes to the herring.

iv) Calculate the amount of energy that enters the food chain per year if the plankton use 0.1% of the available energy.

v) Explain two ways in which energy is lost in the transfer from the copepod to the herring.

c) Herring is rich in protein and lipid (fats and oils), and has a high vitamin D content. Describe the uses of each of these food substances in a balanced human diet.

5 Apart from water, bread contains mainly starch, protein and lipid. Imagine a piece of bread about to start its journey through the human gut.

a) Copy and complete the table to show the digestive processes that happen to the bread as it passes through the mouth, stomach and duodenum (the first part of the small intestine).

Part of gut	Enzyme(s) secreted	Action on bread	Other functions
mouth	amylase		saliva moistens food, teeth mechanically break down food
stomach		digests protein into peptides	muscular churning mixes bread with enzyme
duodenum (enzymes from pancreas)	amylase		–
	trypsin		–
		digests lipid into fatty acids and glycerol	–
duodenum (bile from liver)	–		bile is alkaline to help neutralise stomach acids

b) Explain the ways in which the last part of the small intestine (the ileum) is adapted for absorbing the breakdown products from the digestion of the bread.

c) Explain, with the help of a labelled diagram, how the contents of the gut are moved along by the process of peristalsis.

d) The disease **cholera** is caused by a bacterium that lives in the human gut. One symptom of the disease is chronic diarrhoea (loss of water in the faeces), which leads to dehydration and death, particularly in children.

i) Suggest how infection of the gut with the bacteria results in diarrhoea.

ii) Children with diarrhoea caused by diseases such as cholera can be treated using the **oral rehydration method**. Explain how this works.

Index